FUTURE OF
RENEWABLE
ENERGY

The Future of Renewable Energy

What Is the Future of Fossil Fuels?

Hal Marcovitz

ReferencePoint Press®

San Diego, CA

About the Author

Hal Marcovitz is a former newspaper reporter and columnist. He has written more than 150 books for young readers. He lives in Chalfont, Pennsylvania, in a home powered by solar energy.

Cover: Thinkstock.com
Steve Zmina: 8, 15, 21, 28, 34, 41, 46, 54, 59

LIBRARY OF CONGRESS CATALOGING-IN-PUBLICATION DATA

Marcovitz, Hal.
  What is the future of fossil fuels? : by Hal Marcovitz. -- (Future of renewable energy series)
  Includes bibliographical references and index.
  ISBN 978-1-60152-612-0 (hardback)
  ISBN 1-60152-612-1 (hardback)
  1. Fossil fuels. I. Title.
  TP318.M37 2013
  553.2--dc23
                                                                2013029017

Contents

Foreword

What are the long-term prospects for renewable energy?

In his 2011 State of the Union address, President Barack Obama set an ambitious goal for the United States: to generate 80 percent of its electricity from clean energy sources, including renewables such as wind, solar, biomass, and hydropower, by 2035. The president reaffirmed this goal in the March 2011 White House report *Blueprint for a Secure Energy Future.* The report emphasizes the president's view that continued advances in renewable energy are an essential piece of America's energy future. "Beyond our efforts to reduce our dependence on oil," the report states, "we must focus on expanding cleaner sources of electricity, including renewables like wind and solar, as well as clean coal, natural gas, and nuclear power—keeping America on the cutting edge of clean energy technology so that we can build a 21st century clean energy economy and win the future."

Obama's vision of America's energy future is not shared by all. Benjamin Zycher, a visiting scholar at the American Enterprise Institute, a conservative think tank, contends that policies aimed at shifting from conventional to renewable energy sources demonstrate a "disconnect between the rhetoric and the reality." In *Renewable Electricity Generation: Economic Analysis and Outlook* Zycher writes that renewables have inherent limitations that can be overcome only at a very high cost. He states: "Renewable electricity has only a small share of the market, and ongoing developments in the market for competitive fuels . . . make it likely that renewable electricity will continue to face severe constraints in terms of competitiveness for many years to come."

Is Obama's goal of 80 percent clean electricity by 2035 realistic? Expert opinions can be found on both sides of this question and on all of the other issues relating to the debate about what lies ahead for renewable energy. Driven by this reality, *The Future of Renewable Energy* series critically examines the long-term prospects for renewable energy by delving into the topics and opinions that dominate and inform renewable energy policy and debate. The series covers renewables such as solar, wind, biofuels, hydrogen, and hydropower and explores the issues of cost and affordability, impact on the environment, viability as a replacement for fossil fuels, and what role—if any—government should play in renewable energy development. Pointed questions (such as "Can Solar Power Ever Replace Fossil Fuels?" or "Should Government Play a Role in Developing Biofuels?") frame the discussion and support inquiry-based learning. The pro/con format of the series encourages critical analysis of the topics and opinions that shape the debate. Discussion in each book is supported by current and relevant facts and illustrations, quotes from experts, and real-world examples and anecdotes. Additionally, all titles include a list of useful facts, organizations to contact for further information, and other helpful sources for further reading and research.

Visions of the Future: Fossil Fuels

According to the Old Testament, the Valley of Elah is where David slew Goliath. The valley is in Israel, not far from Jerusalem. Today geologists spend a lot of time in the valley. They are not looking for artifacts of an ancient civilization but, rather, deposits of shale rock laced with a substance known as kerogen. Shale is a sedimentary rock, meaning it formed through the coalescence of mud, sand, salt, and other minerals found in the earth.

Kerogen is rich in hydrocarbons, the same chemicals found in oil, natural gas, and coal—the three fossil fuels. Unlike oil, which is extracted from wells drilled as much as 7 miles (11.3 km) deep, kerogen is found much closer to the surface. In Israel geologists have determined the kerogen deposits are buried no more than 1,000 feet (305 m) below the surface. If the deposits of the hydrocarbon-rich mineral had been buried deeper, the intense pressure found at those depths would have, over the course of millions of years, turned the substance into oil.

The kerogen deposits were discovered in 1980, but at that time the technology to extract the substance—also known as oil shale—did not exist. The oil industry has since developed methods to draw kerogen out of the shale and refine it as crude oil is refined—into gasoline, heating oil, and other petroleum products. Essentially, hundreds of tiny holes are drilled into a field; steel cables are fed into the holes and heated. This transfer of heat causes the kerogen to rise to the surface. "We think that within a decade we can get 50,000 to 100,000 barrels a day,"[1] says Relik Shafir, chief executive officer of Israel Energy Initiatives, which plans to extract oil shale from the Valley of Elah.

Thirst for Fossil Fuels

Unlike its oil-rich neighbors in the Middle East, Israel has never had an oil industry. Between 1948 and 1986, when the search for oil was finally abandoned, Israeli geologists sunk 440 test wells—all of them dry. But now, if Israel is able to exploit its oil shale reserves, the country has the potential to become a major international energy supplier. According to Israel Energy Initiatives, the amount of kerogen available beneath the ground in Israel is equivalent to some 250 billion barrels of crude oil—roughly the amount of crude found beneath the desert of Saudi Arabia, one of the most oil-rich countries on earth.

Israel's oil shale reserves would help slake the enormous thirst for fossil fuels that exists in the world today. Each year, the world's users of petroleum—predominantly motor vehicle drivers, as well as home-owners and business owners who heat with oil—burn through some 1.6 billion barrels of crude oil, according to 2011 statistics compiled by BP, one of the world's largest oil companies. (There are 42 gallons [159 L] to a barrel of crude, which is unrefined oil.) BP compiled statistics for other fossil fuels as well. Coal users—mostly electric companies that use the mineral to fuel their power-generating plants—consume 3.7 billion tons (3.4 billion metric tons) a year. The users of natural gas, which is employed mainly as a heating fuel, burn through some 3.2 trillion cubic feet (90.6 billion cu. m) a year.

Since the 1700s, when coal replaced wood as a fuel for cooking, heating, and the earliest industries, fossil fuels have been the chief source of power in the global society. Nuclear power plants started making electric power in the 1950s, and renewable energy sources—including solar and wind power and biofuels—have gained in popularity. But all those alternative sources of energy combined are easily outpaced by the worldwide consumption of fossil fuels. According to the Paris-based International Energy Agency, which monitors worldwide energy use, in 2010, oil, natural gas, and coal provided 71.4 percent of the world's energy needs. Says Robert Bryce, editor of the trade journal *Energy Tribune*, "The overwhelming majority of the power we use comes from hydrocarbons because they can provide us with the reliable and abundant power we desire."[2]

Top Oil-Producing Nations

The fossil fuels oil, coal, and natural gas supply the largest share of energy needs worldwide; in part, this is because fossil fuels can be found on land and in oceans all across the globe. The world's top oil-producing nations, for example, are located on five continents. This map shows total oil produced in barrels per day (bbl/day) by these nations.

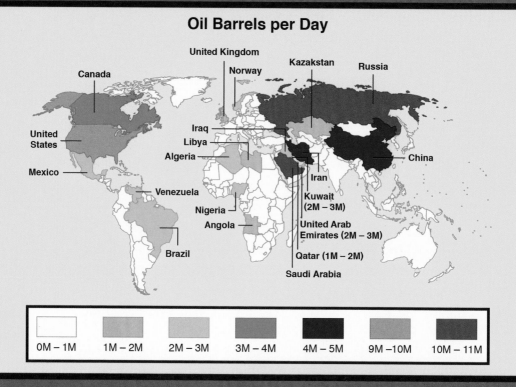

Oil Barrels per Day

Legend: 0M – 1M | 1M – 2M | 2M – 3M | 3M – 4M | 4M – 5M | 9M –10M | 10M – 11M

Source: US Energy Information Administration, "World Regions Oil Production Oil Consumption Proved Reserves," 2011. www.eia.gov.

The Carboniferous Period

Fossil fuels are extracted from the earth, where they formed some 286 million to 360 million years ago from the remains of animals and plants—hence the name "fossil fuels." This is the era known as the Carboniferous period, which gets its name from carbon. Carbon is the basic element found in fossil fuels.

During the Carboniferous period, the earth was covered with swamps and bogs. When living things died, they decomposed and sank to the bottom of the swamps, forming a spongy substance known as peat. Over millions of years, layers of mud, rock, and sand sank deep into the swamps, burying the peat. Eventually, the peat was covered by hundreds and even thousands of feet of earth. Over a large portion of the earth, seawaters covered the deposits of peat as well. The enormous weight of the mud, rock, and sand atop the peat compressed the substance into coal, oil, and natural gas.

There is virtually no area of the planet where fossil fuels are not extracted from the ground. In Oklahoma, Texas, and other states, derricks pump oil from wells twenty-four hours a day, seven days a week. In the Gulf of Mexico and North Sea off the coast of Denmark, drilling platforms stand atop thousands of feet of seawater. Oil is extracted from wells above the Arctic Circle and piped across thousands of miles of land in Alaska, Canada, and Russia. In the Middle East, oil deposits in such places as Saudi Arabia and Kuwait have made those countries among the wealthiest in the world. Coal mining is found on virtually every continent. It is dug out of mines that tunnel thousands of feet into the earth and stripped from mountaintops where the mineral can be found relatively near the surface. Natural gas is abundant as well. One of the richest deposits of natural gas on the planet can be found in a region known as the Marcellus Shale, which covers 54,000 square miles (139,859 sq. km) in parts of New York, Pennsylvania, Ohio, West Virginia, Maryland, Kentucky, Tennessee, and Virginia. The US Geological Survey estimates the Marcellus Shale contains some 30.7 trillion cubic feet (869 billion cu. m) of natural gas.

Peak Oil

Those statistics would seem to suggest that automobile owners, electric power plant operators, and homeowners will never find themselves lacking a source of fossil fuels. However, many experts fear fossil fuel producers may soon draw out a majority of the world's attainable supply of oil, coal, and natural gas. Says Sascha Müller-Kraenner, senior policy adviser to the environmental group Nature Conservancy, "A ghost haunts the energy industry. Its name is 'peak oil' and it prophesizes the end of the oil era—or at least the end of a cheap, seemingly inexhaustible lubricant for the global economy."[3] In fact, for some of the world's most reliable oil suppliers, the end may be relatively near. BP has estimated that the Middle East nation of Oman, with some 6 billion barrels in reserve, will likely run out of oil by 2029. Even Saudi Arabia, which sits atop an estimated 264 billion barrels of oil, may have reached its peak output. In 2012, the US government released a report suggesting the Saudis may have overestimated their reserves by 40 percent. And Citibank, the international banking firm, suggested the Saudis may run out of oil as soon as 2030.

Moreover, energy industry analysts point out that the last major crude oil discovery was in 1969, when vast reserves were found off the shores of Alaska in Prudhoe Bay, an inlet of the Beaufort Sea. In the meantime, consumption has risen, forcing the oil producers to meet demand by tapping deeper into existing sources.

To stave off the realities of peak oil and the depletion of the other fossil fuels, energy companies have turned to technology. The process that will be employed in Israel to lift the kerogen out of the earth is one example. Another is hydraulic fracturing—commonly known as "fracking"—in which tons of water and chemicals are injected under high pressure into shale, forcing the expulsion of natural gas. That is the process that is widely used to extract natural gas in the Marcellus Shale region.

A Troubling Future?

But such efforts do come with a heavy price. Fracking is believed to have significant environmental impacts—the water left over after a fracking operation is regarded by environmental activists as a source of pollution. And the use of all fossil fuels is believed to contribute to climate change—a warming of the earth's surface caused by the expulsion of carbon into the atmosphere when fossil fuels are burned.

Today oil, coal, and natural gas provide reliable sources of energy, but many experts believe those supplies are hardly inexhaustible. Meanwhile, the very real danger of climate change hovers over the entire planet, prompting many critics to suggest that the continued use of fossil fuels could have a devastating impact on the future of human society.

Chapter One

Are Fossil Fuels Affordable?

Fossil Fuels Are Affordable

Fossil fuels are an established source of energy for consumers. Oil, coal, and natural gas remain in ample supply, and in fact, new sources of natural gas discovered in America are expected to help make the country energy independent. New technologies help keep fossil fuels affordable—cars, for example, are far more fuel efficient than they were a generation ago. And unlike most renewable energy sources, the infrastructure required to offer fossil fuels to consumers is well established, with pipelines, supertankers, refineries, mines, drilling platforms, and other modes of supply and delivery already in place.

The Debate

Fossil Fuels Are Too Costly

Fossil fuels may be in abundance, but their extraction can often be an expensive enterprise. In no place is this more evident than above the Arctic Circle, where energy companies have flocked because new reserves have been discovered there. Establishing drilling platforms in this most inhospitable of climates is expensive and has yet to pay off for energy companies. Meanwhile, skeptics believe the cost of fracking will outweigh the benefits of withdrawing shale gas, and many people agree human civilization is facing tremendous economic consequences due to fossil fuel emissions.

Fossil Fuels Are Affordable

"We use hydrocarbons—coal, oil, and natural gas—not because we like them, but because they produce lots of heat energy, from small spaces, at prices we can afford, and in quantities that we demand."

—Robert Bryce, *Power Hungry: The Myths of Green Energy and the Real Fuels of the Future*. New York: PublicAffairs, 2010, p. 4.

Robert Bryce is editor of the trade journal *Energy Tribune*.

Fossil fuels are affordable because they are abundant. According to BP, worldwide reserves of oil, natural gas, and coal available for extraction and delivery to the marketplace include 1.6 trillion barrels of oil, 860 billion tons (780 billion metric tons) of coal, and 7.4 quadrillion cubic feet (210 trillion cu. m) of natural gas. Experts disagree on how long those reserves will last, but many believe there will be enough fossil fuels to feed the world energy habit at least into the twenty-second century and perhaps indefinitely. These experts point out that new discoveries of fossil fuel reserves and new technologies designed to extract fossil fuels from depths previously believed to be unreachable are now available. Says Nigel Lawson, former secretary of energy for Great Britain, "The so-called 'peak oil' theory, which suggests that within the foreseeable future the world will run out of fossil fuels—coal, oil and gas—has never looked more absurd."[4]

The infrastructure to deliver those supplies to consumers has been in place for decades. Coal mining is a well-established industry in West Virginia, Kentucky, Pennsylvania, and other states. Thousands of miles of natural gas pipelines crisscross continents, providing the fuel to homes and businesses. Oil pipelines also cross huge expanses of land—the 4-foot-wide (1.2 m) Trans-Alaska oil pipeline, for example, started carrying oil in 1977; it delivers crude over some 800 miles (1,287 km) of Alaskan wilderness from the Arctic Circle to the port city of Valdez,

Alaska. In Valdez, "supertankers" receive the fuel and then embark for international destinations—usually port cities where refineries turn the crude into gasoline. There are already 570 supertankers in operation in the world. These ships, called very large crude carriers, typically hold 2 million barrels of crude. With so much infrastructure already in place, it means few new pipelines, refineries, and tankers have to be built to deliver fossil fuels, a factor that helps keep prices affordable.

Technology Improves Efficiency

Advances in technology have made the use of fossil fuels more efficient; therefore, a barrel of crude oil or ton of coal produce more energy than they did years ago. A simple law of economics holds that demand drives prices; when demand is held in check, prices do not rise. Cars provide a typical example of how improvements in technology have played a key role in keeping demand for oil in check. According to the Washington, DC–based Pew Environmental Group, which studies environmental issues, in 1975 the average American car was able to travel less than 15 miles (24 km) on a gallon of gas. In 2010, after American auto companies spent thirty-five years developing more fuel-efficient engines and lighter cars, fuel efficiency for the average car rose to nearly 35 miles (56 km) per gallon of gas. Therefore, people need less gasoline to drive the same dis-tances, a factor that has helped sustain the glut of crude on the market.

> "The so-called 'peak oil' theory, which suggests that within the foreseeable future the world will run out of fossil fuels—coal, oil and gas—has never looked more absurd."[4]
>
> —Nigel Lawson, former secretary of energy for Great Britain.

Says Addison Armstrong, director of research at Tradition Energy, a Stamford, Connecticut–based firm that invests in energy companies, "There's so much spare capacity right now it's very difficult to see prices much higher."[5] Moreover, the auto industry is under pressure to make cars even more fuel efficient. In 2012 the US Department of Energy set regulations mandating new cars average 55 miles (89 km) per gallon by 2025.

Coal-fired electrical plants have also found ways to become more efficient. Coal plants make electricity by heating water into pressurized steam, which runs turbines that generate electricity. The Sante Fe Institute, a New Mexico–based organization that examines public policy issues, has estimated that in the 1880s most coal-fired plants operated at an efficiency rate of a mere 3 percent—meaning they wasted 97 percent of the energy they produced. Today coal plants run on an efficiency level of about 70 percent. When energy plants produce power more efficiently, less power is wasted. Such efficiencies are reflected in lower prices paid by consumers.

For most electric power companies, the main alternative to coal is nuclear fuel—but nuclear energy is far more expensive than coal. The chief reason electricity produced with coal is cheaper than electricity produced with nuclear power is the tremendous cost of constructing nuclear power plants. The Cambridge, Massachusetts–based energy trade research group Synapse Energy Economics estimates the cost of building a new nuclear power plant at $18 billion. Construction of nuclear plants is very expensive because they require huge amounts of infrastructure—massive containment buildings that can withstand earthquakes and other natural disasters, ensuring the dangerous nuclear fuel inside is not exposed to the outside world. Certainly, electricity consumers who live in a city served by a new nuclear plant would see the cost reflected in their monthly electric bills. A new coal-fired plant is much cheaper. Synapse has estimated the construction cost for a coal-fired plant at $3 billion.

Energy Independence

New discoveries such as natural gas in the Marcellus Shale will help keep fossil fuels affordable. According to the US Energy Information Administration, since 2000 the discovery of shale gas reserves in America has boosted the amount of natural gas available in the country by 60 percent. In 2000 about 1.5 quadrillion cubic feet (42.5 trillion cu. m) of gas was believed to be available beneath the ground in the continental United States; that estimate was revised upward in 2011 to more than 2.5 quadrillion cubic feet (70.8 trillion cu. m). Says *Energy Tribune* editor Robert Bryce, "It's abundantly clear that the United States has enormous quan-

New Technology Keeps Gasoline Prices Low

In 1975, most cars and trucks were able to travel less than 15 miles on a gallon of gasoline; by 2010, after thirty-five years of improving engine efficiency and making vehicles lighter, the average car can travel 35 miles on the same gallon of gas while small trucks can travel about 25 miles on a gallon of gas. Better fuel efficiency means people use less gasoline. Less demand helps keep oil prices low.

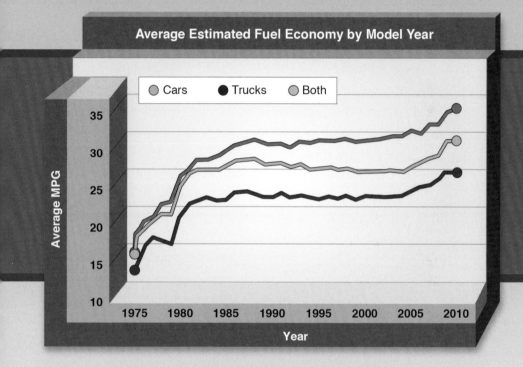

Source: Pew Environmental Center, "History of Fuel Economy," April 4, 2011. www.pewenvironment.org.

tities of natural gas—more gas than ever thought possible."[6] Abundant supply will likely keep future costs low.

Moreover, much of the shale gas available in America is located near population centers—meaning energy companies do not have to pay the cost of shipping or piping the gas over hundreds or even thousands of miles to big cities. The proximity of shale gas reserves to big cities is another factor that should help keep prices low.

The abundance and proximity of shale gas has led the Energy Information Administration to predict America could be energy independent—meaning the country would not have to rely on foreign fossil fuel producers—by 2030 and even become an energy exporter by 2035. Says Thanassis Cambanis, a research fellow at the Century Foundation, a Washington, DC–based group that examines international economic issues, "Given the United States' huge appetite for fuel, energy independence has always seemed more of a dream than a realistic prospect. But today . . . energy independence is starting to loom in sight."[7]

> "Given the United States' huge appetite for fuel, energy independence has always seemed more of a dream than a realistic prospect. But today . . . energy independence is starting to loom in sight."[7]
>
> —Thanassis Cambanis, a research fellow at the Century Foundation.

If the day does come when America can declare its energy independence, it is likely that fossil fuel prices will remain stable for years. Today America imports nearly 13 percent of its oil from Persian Gulf countries, according to 2012 statistics compiled by the Energy Information Administration. (Although 13 percent may seem like a small slice of overall oil consumption, suddenly slashing those imports would throw the American energy market into chaos, causing steep demand and a huge spike in prices.)

The Persian Gulf is one of the most volatile regions on the planet, an area rife with religious extremists, terrorists, and governments hostile to American interests. Whenever conflict breaks out in the Persian Gulf, the interruption of supply from the region usually results in a hike in oil prices. No example better illustrates this fact than the Iraq War, which erupted in 2003 when the American military led an international coalition to oust Iraqi dictator Saddam Hussein. When Iraq was invaded, crude oil sold for $25 a barrel on world markets. By late 2011, when most combat troops were withdrawn from Iraq, crude sold for about $112 a barrel—a near fivefold increase. Years of warfare disrupted supply and threw uncertainty into the international fuel markets, causing prices to skyrocket. By using oil produced within

American borders only, consumers would not be exposed to the steep hikes in oil prices that occur during periods of warfare in foreign oil-producing regions.

An Integral Part of the Economy

Americans would enjoy stable fuel prices if all the fossil fuels they needed were produced within US borders. Among the factors that would keep fossil fuel prices low is the discovery of new reserves of natural gas and, along with it, energy independence. Americans can also look forward to continued improvements in technology that affect how people use fossil fuels—such as more fuel-efficient automobiles. Moreover, infrastructure that is already in place can tap into what are regarded as vast reserves of oil, coal, and natural gas, helping ensure that fossil fuels remain affordable and an integral part of US and world economies far into the future.

Fossil Fuels Are Too Costly

"We are not running out of oil, but we are running out of oil that can be produced easily and cheaply."

—David King and James Murray, "Climate Policy: Oil's Tipping Point Has Passed," *Nature*, January 26, 2012, p. 433.

David King is director of Smith School of Enterprise and the Environment at the University of Oxford, and James Murray is a University of Washington oceanographer.

Despite the abundance of fossil fuels, their extraction and delivery to consumers remain an expensive undertaking reflected in the prices people pay for gasoline, heating oil, coal, and natural gas. Often contributing to the expense of supplying fossil fuels is the uncertainty of the weather, which can play a role in the rise of energy prices. When Hurricane Isaac made landfall in August 2012, gasoline prices rose an average of seven cents per gallon at the pump in just two days. As the hurricane swept through the Gulf Coast states, eleven refineries were forced to shut down. Those refineries produce nearly 3 million barrels of gasoline a day, or about 12 percent of all the gasoline refined from the oil withdrawn by the offshore platforms in the Gulf of Mexico. "This week, the main factor that is driving up gas prices is the landfall of Hurricane Isaac in the Gulf Coast," Michael Green, a spokesperson for the American Automobile Association, said at the time. "Whenever you see supplies drop, prices increase."[8]

Hurricanes cause only a temporary rise in prices; after most major storms, refineries reopen and the cost of gasoline usually returns to prestorm prices. But energy companies are looking for new sources of fossil fuels—often in places where the weather never improves. Many of the world's untapped reserves of fossil fuels are located well above the Arctic Circle. Searching for oil and natural gas in subzero conditions has placed incredible financial risks on energy companies, and these risks are sure to be reflected in the prices consumers pay for fossil fuels.

Mishaps in the Arctic

The US Geological Survey estimates the region above the Arctic Circle holds 90 billion barrels of recoverable oil and 1.6 trillion cubic feet (45 billion cu. m) of natural gas. Royal Dutch Shell has already spent $5 billion in attempts to establish drilling platforms above the Arctic Circle but has yet to extract a single drop of oil from its Arctic operation. A major setback occurred in January 2013 when Shell's *Kulluk* oil rig blew into rocks during a storm and had to be towed back to shore in a huge rescue operation that required the use of fifteen vessels, two helicopters, and the efforts of more than 730 people.

And it is likely that Shell's investment in its Arctic drilling program has only just begun. According to the Energy Information Administration, it could take as long as ten years for a well to begin producing oil once the drilling process begins. That means companies like Royal Dutch Shell have to maintain expensive Arctic drilling platforms for a decade or more before they begin producing crude.

The mishap involving the *Kulluk* rig is not the only problem Shell has experienced in northern climates. Shell has also used an ocean-based drilling platform aboard the ship *Noble Discoverer* to extract oil from under the seas near Alaska. Twice in 2012 the *Noble Discoverer* suffered mishaps—in July the ship nearly ran aground while sailing through Alaska's Aleutian Islands; four months later the ship stalled in the water when a fire broke out aboard the vessel. Finally, the *Noble Discoverer*—which Shell leases from a Swiss company at a cost of nearly $2 million a week—had to return to port in Seattle for repairs.

Environmental advocates suggest that mishaps similar to the *Kulluk* and *Noble Discoverer* incidents will become common in the Arctic and similar regions because such areas are simply too inhospitable to maintain safe and therefore profitable drilling operations. In 2013 the international accounting agency Ernst & Young issued a report affirming the view that such ventures carry very high risks. The report also suggests that the high cost of drilling in the Arctic would eventually be passed on to consumers, meaning motorists will pay higher prices at the pump. Says the Ernst & Young report:

> The quest for Arctic oil and gas resources is not for the faint of heart nor for those with less-than-deep pockets. Rather, Arctic oil and gas resource development is both high-cost and high risk. . . . These operations are clearly on the outer limits of both safety and commercial viability for the industry and a spill or accident there would be catastrophic. The economics of Arctic development are also looking forward to even higher oil prices.[9]

False Promise of Shale Gas

Even when new sources of fossil fuels are located closer to home, it still takes a tremendous investment to begin new drilling or mining operations. Indeed, for all the promise that shale gas holds for delivering energy to consumers, extraction of the resource is an expensive undertaking. A single well drilled for hydraulic fracturing can cost as much as $4 million. What's more, natural gas companies must lease the properties from the owners to establish the wells. In southern Illinois alone, fracking companies have already spent $250 million on leases. With fracking operations virtually coast to coast, lease costs can easily amount to billions of dollars.

> "Arctic oil and gas resource development is both high-cost and high risk. . . . These operations are clearly on the outer limits of both safety and commercial viability for the industry and a spill or accident there would be catastrophic."[9]
>
> —Ernst & Young, an international accounting agency.

Given those numbers, the economic investment in hydraulic fracturing is expected to be enormous. According to the Interstate Natural Gas Association of America, there were thirty-one thousand hydraulic fracturing

Natural gas prices spiked in the early 2000s, but after hydraulic fracturing found vast new reserves of the fuel, the prices dropped. Hydraulic fracturing, also known as fracking, is a method of extracting gas from the ground by forcing it to the surface using pressurized water and chemicals. Even though fracking is expected to become a major source of natural gas, the US Energy Information Administration (EIA) forecasts that by 2035 prices may nearly double due to the high cost of establishing hydraulic fracturing wells. In 2035, the EIA projects it will cost consumers $7.37 per million BTUs, or British thermal units.

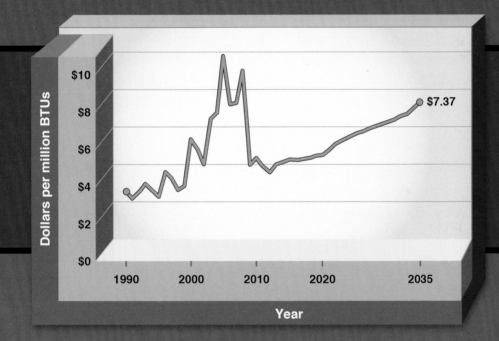

Note: A BTU is a measurement of energy; the energy in 1 million BTUs is roughly equal to the energy in about 1,000 cubic feet of natural gas. Most American homes burn through 70,500 cubic feet of natural gas a year.

Source: US Energy Information Administration, "Annual Energy Outlook 2012," June 2012. www.eia.gov.

wells in operation in America in 2007; by 2020 that number is expected to grow to three hundred thousand.

The costs of leasing land and setting up extraction operations are sure to be reflected in the prices consumers will pay to heat their homes.

According to the International Monetary Fund, a Washington, DC–based institution that forecasts economic trends, the price of natural gas in America is expected to rise 38 percent between 2013 and 2020. The American Gas Association reports that the average American home that heats with gas burns through 70,500 cubic feet (1,996 cu. m) a year at a cost of about $550 in 2013. A 38 percent hike by 2020 would raise that cost to about $760 a year.

Moreover, those numbers are based on the industry's projections of the quantity of gas that will be made available to consumers. Those projections may be too rosy, however. With less natural gas in reserve for the future, existing supplies will cost more. Deborah Rogers is founder of Energy Policy Forum, which has studied the rise of the fracking industry. She says industry projections that suggest hydraulic fracturing can meet America's natural gas demand for the next century include the use of gas that is still too deep for current fracking methods to withdraw. Says Rogers, "While the resource may exist underground, it may not be readily extractable because it is not technologically possible or economically viable in present-day terms. In reality, reserves—gas that can be extracted right now and valued in the market—are more like 11 years' worth."[10]

Hidden Costs of Fossil Fuels

A hidden cost of fossil fuels is the impact oil, coal, and natural gas have on the health of Americans. The National Academy of Sciences estimates that fossil fuel pollution adds $120 billion a year to the cost of delivering health care in America. These pollutants include particles of soot and the chemical nitrogen oxide, which can cause lung damage. According to the academy, twenty thousand people a year die from respiratory diseases caused by fossil fuel pollution, and hundreds of thousands of others receive medical care due to lung ailments caused by fossil fuel pollution.

Another hidden cost is the damage to buildings, streets, and bridges expected to be caused by climate change that is occurring due to the use of fossil fuels. Some cities are already taking steps to protect their infrastructure against rising sea levels—an expected outcome of climate change. In 2013 New York City mayor Michael Bloomberg announced a $20 billion plan to build flood walls and levees along the city's 520 miles (837 km) of coastline. Bloomberg has also proposed providing aid to oceanfront and riverfront property owners to help them raise the foundations of their buildings so they can endure the flooding that is expected to become routine as the earth's surface warms. "This is urgent work and it must begin now," Bloomberg said when he announced the plan. "If we're going to save lives and protect the lives of communities, we're going to have to live with some new realities."[11]

Fossil fuels are abundant, but their abundance does not necessarily mean they can be delivered cheaply, safely, and efficiently. Their continued use could ultimately lead to high costs as energy companies invest billions of dollars in new extraction technologies and locations. Finally, as the by-products of fossil fuels enter the atmosphere and cause pollution, as well as climate change, communities like New York City may find themselves spending billions of dollars to protect themselves against rising sea levels. And even people who do not live near the coasts may find their medical bills rising as they deal with the health effects of pollution caused by fossil fuels.

> "While the resource may exist underground, it may not be readily extractable because it is not technologically possible or economically viable in present-day terms."[10]
>
> —Deborah Rogers, founder of Energy Policy Forum.

Chapter Two

Can Fossil Fuels Be Compatible with the Environment?

Fossil Fuels Can Be Compatible with the Environment

Through carbon capture technology, fossil fuels can be made to burn more cleanly, which minimizes the effect of oil, coal, and natural gas on the environment. Moreover, research is underway to make wider use of natural gas—the cleanest of the fossil fuels. Through the use of compressed natural gas, automobiles and power plants can be made to run virtually emission free. Even as these efforts proceed, a faction of the scientific community remains steadfast in its belief that fossil fuels are not responsible for climate change.

The Debate

Fossil Fuels Are Not Compatible with the Environment

The Clean Air Act may have helped reduce toxic emissions from coal-fired factories, but the fracking boom has posed new threats to the environment as toxic chemicals are pumped into the ground under high pressure. Accidents such as the 2010 explosion on the *Deepwater Horizon* drilling rig could have adverse impacts on the environment for decades. Meanwhile, a strong body of evidence has linked greenhouse gases to climate change. Composed mostly of carbon dioxide, such gases trap heat in the earth's atmosphere and cause global warming.

Fossil Fuels Can Be Compatible with the Environment

"New coal plants are best-in-class global leaders in generating efficient, clean, reliable and affordable electricity."

—Quoted in National Mining Association, "NMA's Quinn Warns Against Harm to Jobs and US Economy," June 25, 2013. www.nma.org.

Hal Quinn is president of the National Mining Association.

Natural gas is the cleanest burning of the fossil fuels. According to the US Environmental Protection Agency, when natural gas is burned it emits half as much carbon dioxide as coal. Therefore, with hydraulic fracturing producing a greater share of energy in the future, experts believe the environmental impact of fossil fuels can be minimized.

One technological development that is sure to make fossil fuels more environmentally friendly is the conversion of coal-burning electric generating stations and oil-burning motor vehicles to natural gas power. These conversions can be accomplished through the use of compressed natural gas, or CNG, technology, in which natural gas is compressed to 1 percent of its original volume and stored in cylinders. Ignition of the released CNG fuels the engine. In 2013 America's Natural Gas Alliance, a consortium of energy companies, underwrote a project to convert a half dozen factory-produced cars, including a BMW sport-utility vehicle and a Ford Mustang, into CNG vehicles to show the viability of the technology. Says Peter Voser, chief executive officer of Royal Dutch Shell, "The world needs to follow America's lead and take full advantage of the cleanest-burning fossil fuel, and that's natural gas. Increased use of natural gas is the biggest single step that the world can take today to begin reducing [carbon dioxide emissions]."[12]

It is expected to be many years before CNG vehicles become economically feasible. When they go into mass production, the average price for a CNG car is expected to be $11,000 more than a standard auto. Nevertheless, proponents of natural gas are starting to make the case for conversion to CNG. In 2013 Philadelphia Gas Works, the utility that provides gas to city residents, approved spending $438,000 to buy CNG-powered vehicles for gas company maintenance workers.

Capturing Carbon

CNG may hold a lot of promise as an environmentally friendly fossil fuel. Some scientists suggest it is also possible to take fossil fuels in their existing forms and scrub them of their carbon emissions. Currently, most research in the field is aimed at scrubbing carbon emissions created through the burning of coal.

> "The world needs to follow America's lead and take full advantage of the cleanest-burning fossil fuel, and that's natural gas."[12]
>
> —Peter Voser, chief executive officer of Royal Dutch Shell.

Climate scientists report that when it comes to carbon emissions, coal is a major culprit: According to the Arlington, Virginia–based Center for Climate and Energy Solutions, coal provides 27 percent of the energy used worldwide yet accounts for 43 percent of carbon emissions. But scientists are developing methods for scrubbing coal-fired emissions of their carbon dioxide content before they are released into the atmosphere. This clean coal technology, as it is now known, offers great promise for the future.

Most of the research has focused on carbon capture, also known as carbon sequestration, a process in which the carbon gas is captured before it is emitted through smokestacks. During this process the carbon is concentrated, then pumped into the ground in natural geologic formations for permanent storage. Carbon capture technology research focuses mostly on a postcombustion process in which the carbon is removed from the flue gases—the emissions that are vented out of the smokestacks at coal-fired power plants. To capture the carbon, the flue gases are filtered through chemical solvents that absorb the carbon. This is the capture phase of the project. According to the International Energy Agency, carbon capture technology can remove more than 85 percent of greenhouse gases from fossil fuel emissions.

Technology Is Moving Forward

The captured carbon is then transported through a pipeline to fissures in the earth—some beneath the seafloor—where it can be indefinitely stored. According to the Intergovernmental Panel on Climate Change (IPCC), an agency established by the United Nations and the World Meteorological Organization, there may be room beneath the ground for as much as 10 trillion tons (9 trillion metric tons) of carbon dioxide. Indeed, the IPCC estimates that enough space exists beneath the surface of the earth to store carbon for the next one thousand years. Says Sally Benson, executive director of the Stanford University–based Global Climate and Energy Project, "The goal of carbon sequestration is to permanently store the carbon dioxide. Permanent meaning very, very long-term, geological time periods."[13]

Carbon capture technology is still in its infancy. By 2013 there was just a single coal-burning power plant under construction in America that would feature carbon capture technology. That plant is located in Kemper County, Mississippi. Other carbon sequestration power plants in the planning stages are located in Ector County, Texas; Houston, Texas; Kern County, California; and Meredosia, Illinois.

Climate Change Skeptics

Even as research into CNG and carbon capture technology continues, many energy company executives as well as scientists continue to maintain there is no direct connection between carbon emissions and global warming. The greenhouse effect first surfaced as a concern in 1988 when James E. Hansen, a climate scientist for the National Aeronautics and Space Administration (NASA), testified before a US Senate hearing on global warming. Hansen testified that decades of monitoring atmospheric conditions led him and many other scientists to conclude that when

Increased Use of Natural Gas Good for Environment

Coal, oil, and natural gas each emit greenhouse gases, but natural gas is the cleanest burning of the three fossil fuels, emitting less carbon dioxide than either coal or oil, according to 2011 statistics compiled by the US Energy Information Administration. With natural gas expected to become more of a factor in American energy use, largely supplanting coal and oil, fossil fuel supporters argue that the continued use of fossil fuels would have a minimal impact on climate change.

US Energy-Related Carbon Dioxide Emissions by Major Fuel Type, 2011

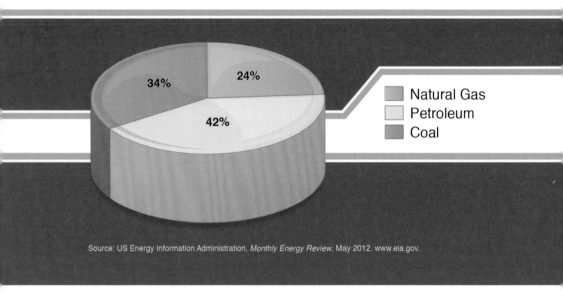

Source: US Energy Information Administration, *Monthly Energy Review*, May 2012. www.eia.gov.

fossil fuels are burned, they emit heat-trapping carbon dioxide gas into the atmosphere. Hansen and others suggest that even a change of just a few degrees in the earth's temperature can lead to the melting of polar ice, flooding in coastal cities, species loss, massive and destructive storms, long-term droughts, and other ill effects.

One of the most renowned skeptics of this theory is Ian Plimer, an Australian geologist, who argues that over the history of the planet the climate has heated and cooled on its own. Plimer argues that the climate change alarmists point out how the earth's temperature has risen a few degrees since the use of fossil fuels became widespread during the Industrial Revolution—an era that is just some two centuries old. To gain a truer picture, Plimer argues, people must study changes in the earth's climate over hundreds of millions of years.

Little Ice Ages

In fact, one theory holds the earth warms and cools because of a periodic wobbling of the planet as it orbits around the sun—an event that has been found to occur roughly every twenty-four thousand to seventy-two thousand years. (This is known as the Milankovitch cycle, named after Serbian astrophysicist Milutin Milankovitch, who suggested the wobbles affect the amount of sunlight striking the earth, causing the earth to heat and cool.) Says Andrew J. Waskey, professor of social sciences at Dalton State College in Georgia, "The slight changes in orbit and the tilt of the Earth in relationship to the sun's rays means more or less sunlight hitting the Earth, increasing or reducing the amount of energy available to warm the Earth and its atmosphere."[14]

> "The goal of carbon sequestration is to permanently store the carbon dioxide. Permanent meaning very, very long-term, geological time periods."[13]
>
> —Sally Benson, executive director of the Stanford University–based Global Climate and Energy Project.

Plimer adds that in the midst of the Milankovitch cycles, the earth has heated and cooled many times on its own. He suggests that changes in climate are cyclical and random. Moreover, what he calls little ice ages last no more than a few hundred years and occur often—in fact, he says the earth is emerging from the latest of its many little ice ages. He explains, "If we just take the last 2,000 years . . . the planet was hot in Roman and Greek times. Then it cooled in the dark ages, then it warmed in the medieval warmth. Then it cooled in the little ice ages, and . . . we've just come out of the little ice age. Is it any wonder that the planet has warmed up?"[15]

Plimer insists the existence of little ice ages proves the earth warms and cools on its own and the climate is not affected by fossil fuel emissions.

He says, "The hypothesis that human activity can create global warming is extraordinary because it is contrary to validated knowledge from solar physics, astronomy, history, archaeology and geology."[16]

Minimal Impacts

With the development of CNG and carbon capture technologies, many advocates for fossil fuels believe they can be made to burn more cleanly, offering minimal impacts on the environment. Moreover, a faction of the scientific community continues to insist that climate change is a natural process unaffected by the burning of fossil fuels and no amount of fossil fuel emissions have had any effect on the earth's changing climate.

Fossil Fuels Are Not Compatible with the Environment

"The whole business model for the fossil fuel industry is based on burning five times more carbon than is compatible with a livable planet. So what we're saying is, 'Your business model is at war with life on this planet. It's at war with us.'"

—Quoted in *Moyers & Company*, "Hurricanes, Capitalism & Democracy," November 16, 2012. http://billmoyers.com.

Naomi Klein is a journalist and the author of a forthcoming book on climate change.

Old photographs depicting the Industrial Revolution era show factory smokestacks rising high above American cityscapes, each billowing streams of black smoke into the atmosphere. Decades ago, such pictures were regarded as signs of progress—coal-burning factories were busily manufacturing goods, boosting the American economy while providing jobs for millions. Today scientists and political leaders know the truth about what was coming out of those smokestacks—polycyclic aromatic hydrocarbons, or PAHs.

PAHs include hundreds of chemicals found in the fumes of burning coal and oil. The two most common are benzo(a)pyrene and naphthalene. PAHs are responsible for a host of human ailments, including cancer. In 1963 Congress passed the Clean Air Act to ensure noxious chemicals are not released into the atmosphere. The Clean Air Act has helped reduce air pollution, but recent advancements in fossil fuel extraction and use have not only raised new concerns about air quality but water quality as well.

Critics of hydraulic fracturing insist the process of forcing natural gas to the surface causes adverse environmental effects. To make the gas rise, hydraulic fracturing crews inject water and chemicals at high pressure into shale rock. Critics contend the chemicals infiltrate nearby underground reservoirs that provide drinking water. Moreover, fracking has also been found to release toxic gases into the air that are not captured by the drilling companies.

Brain Damage and Bone Cancer

Carol French of Bradford County, Pennsylvania, runs a dairy farm near a fracking site. She says, "Our water changed on March 15, 2011."[17] A few months after the drilling began in 2010, French says, many of her cows grew ill. So did her daughter.

She says, "My daughter was 24 at the time. She had a high fever for three days. I thought she had the flu. She had stabbing pains in her abdomen, and diarrhea."[18] At the hospital, French says, doctors determined her daughter suffered from toxic fluids in her abdomen, spleen, liver, and right ovary. After a few days in the hospital, she recovered and then moved in with a friend. Nine days later, she returned home—and the symptoms returned.

A 2011 study by the Endocrine Disruption Exchange, a Colorado group that examines the medical effects of chemicals in the environment, identified 632 chemicals that are injected into the ground during the fracking process. Among the most toxic chemicals identified by the organization are lead and benzene. Lead is a highly toxic metal that can be particularly devastating to young children, causing brain damage; ingesting benzene can lead to bone cancer.

New fears about air pollution have also been raised since the start of the fracking boom. In 2012 the Endocrine Disruption Exchange tested air quality near fracking sites in Colorado and found high concentrations of methylene chloride, a highly toxic solvent used to clean drilling equipment. According to ShaleTest, a nonprofit Texas-based group that conducts air quality testing near fracking sites, hydraulic fracturing is also responsible for the emissions of at least two other toxic substances: acetone, which can cause liver damage, and carbon tetrachloride, a carcinogen.

Effects Felt for Many Years

In addition to toxic effects on health, fossil fuels can be toxic to the environment. Because of the volatility of fossil fuels, their extractions can be hazardous undertakings that result in serious environmental consequences. No example illustrates this more than the *Deepwater Horizon* explosion in 2010. The explosion on the offshore oil rig in the Gulf of Mexico killed eleven workers and was responsible for widespread environmental damage after the pipeline connecting the rig to the well 35,000 feet (10,668 m) below the surface ruptured. It took BP eighty-five days to cap the well—in the meantime, 5 million barrels of oil leaked into the Gulf of Mexico, fouling the coastlines and killing fish and other marine life.

In 2013 the National Wildlife Federation suggested the environmental impact of the *Deepwater Horizon* spill may endure for many years. According to the federation, the spill killed eight thousand birds, turtles, and marine mammals. The oil seeped into the marshes along the Gulf Coast, where it is likely to remain for years and will be ingested by animals. Says a federation statement, "Though oil is no longer readily visible on the surface, it isn't gone. Scientists have found significant amounts on the Gulf floor, and the oil that has already washed into wetlands and beaches will likely persist for years. We likely will not see the full extent of impacts for many years."[19]

> "My daughter was 24 at the time. She had a high fever for three days. I thought she had the flu. She had stabbing pains in her abdomen, and diarrhea."[18]
>
> —Carol French, a Bradford County, Pennsylvania, dairy farmer whose daughter became ill while living near a fracking site.

Realities of Climate Change

Accidents like the *Deepwater Horizon* explosion are not common. Opponents of fossil fuels nonetheless believe there is daily evidence that supports the argument against the continued use of oil, gas, and coal. That argument rests on the effect those fuels have on climate change. According to the World Meteorological Organization, the weather monitoring agency of the United Nations, 2012 was the ninth-warmest year

Earth's Temperatures Are Rising Because of Fossil Fuels

The level of carbon dioxide in Earth's atmosphere has been steadily rising for decades as a result of the burning of fossil fuels for energy, and, according to NASA scientists, that increase has resulted in a warming planet. The year 2012 ranked as the ninth warmest year since 1880, consistent with a long-term trend of rising global temperatures. And 2012 marked the warmest year on record in the continental United States. Seasonal extremes such as those experienced in the United States during the summer of 2012 are increasing and, according to NASA scientist James E. Hansen, will affect people and other life on the planet.

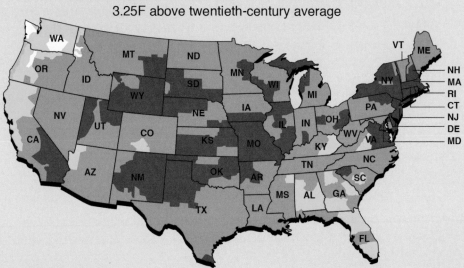

2012 Warmest Year on Record for the Continental US
3.25F above twentieth-century average

Temperature						January–December, 2012
Record Coldest	Much Below Normal	Below Normal	Near Normal	Above Normal	Much Above Normal	Record Warmest

Source: NASA "NASA Finds 2012 Sustained Long-Term Climate Warming Trend," January 15, 2013. www.nasa.gov.

since meteorologists started keeping records in 1850. The average global temperature in 2012 was 58.2°F (14.6°C), or about 1°F (0.6°C) warmer than historical averages.

On its own, 1 degree of temperature would not be regarded as a significant change in the earth's climate, but scientists point out that 2012 marked the continuation of a trend: 2012 was the twenty-seventh consecutive year the global land temperature was above the historical average. Says Gavin Schmidt, a climatologist for NASA's Goddard Institute for Space Studies, "One more year of numbers isn't in itself significant. What matters is this decade is warmer than the last decade, and that decade was warmer than the decade before. The planet is warming. The reason it's warming is because we are pumping increasing amounts of carbon dioxide into the atmosphere."[20]

> "The planet is warming. The reason it's warming is because we are pumping increasing amounts of carbon dioxide into the atmosphere."[20]
>
> —Gavin Schmidt, a climatologist for NASA's Goddard Institute for Space Studies.

The overwhelming opinion of the world scientific community holds that greenhouse gases generated through the burning of oil, coal, and natural gas trap heat in the atmosphere, leading to global warming. Numerous scientific organizations have gone on record endorsing the notion that fossil fuels are at the root of global warming. Typical of these organizations is the American Geophysical Union, which states:

The Earth's climate is now clearly out of balance and is warming. Many components of the climate system—including the temperatures of the atmosphere, land and ocean, the extent of sea ice and mountain glaciers, the sea level, the distribution of precipitation, and the length of seasons—are now changing at rates and in patterns that are not natural and are best explained by the increased atmospheric abundances of greenhouse gases and aerosols generated by human activity during the 20th century.[21]

Infrastructure Tested by Climate Change

According to NASA, climate change will warm the oceans and partially melt glaciers and the polar ice caps, increasing worldwide sea levels. With a smaller environment, species native to cold regions, such as polar bears, will find it difficult to survive. Warmer temperatures will result in evaporation of ocean and stream waters, which will mean more rainfall and severe storms for some places—but arid conditions for others. Again, species will be affected. If forests die due to lack of rainfall or deserts become swampy areas, native species are not likely to adapt to the new conditions and will die off.

This may already be happening in Barrow, Alaska, the northernmost US town. Barrow's roughly forty-five hundred residents report more severe storms than at any time in their memories. The caribou, animals that provide a source of food for residents, are dying off—victims of diseases spread by flies that in years past would never have ventured so far north. Unless greenhouse gas emissions are significantly stemmed, the National Wildlife Federation has predicted that by the end of the twenty-first century, average summer temperatures in the United States could be as much as 11°F (6°C) warmer than they are now.

Despite the existence of the Clean Air Act and similar laws, toxic substances continue to find their way into the environment wherever fossil fuels are extracted and burned. Critics of fossil fuels believe the recent boom in the fracking industry has provided new dangers to the environment. Moreover, the scientific community is virtually united behind the belief that fossil fuel use is responsible for the emission of greenhouse gases and that truly dire consequences await human society unless steps are taken soon to curb the use of coal, oil, and natural gas.

Chapter Three

Can Alternative Energy Take the Place of Fossil Fuels?

Alternative Energy Cannot Take the Place of Fossil Fuels

Alternative energy may be available in abundance, but the world lacks the infrastructure to make wind, solar, and other renewables widely available to industrialized nations. Renewable energy is also unreliable: Wind turbines only produce power when the wind blows; during windless days fossil fuels and nuclear power would be required to provide electricity. Meanwhile, the development of affordable electric cars is still decades away. And many experts believe alternative sources of power simply lack the energy density required to make them effective.

The Debate

Alternative Energy Can Take the Place of Fossil Fuels

There are many examples of how renewable energy has already replaced fossil fuels. In 2013 an experimental aircraft flew cross-country, powered by the energy of the sun. The plane's engineers believe the technology that powered the craft can be applied to other uses. Some countries, among them Scotland and Saudi Arabia, have set goals of being fossil fuel free by 2020, and in the tiny island territory of Tokelau, the fifteen hundred residents already live free of fossil fuels. In more industrialized regions, environmental activists believe conservation could play a role in reducing reliance on fossil fuels.

Alternative Energy Cannot Take the Place of Fossil Fuels

"In two decades, we may achieve greater fuel efficiency with internal-combustion engines, homes may be heated more efficiently, and alternative fuels may have reached the mainstream. But fossil fuels will still power much of the world."

—Torie Bosch, "What Will Turn Us On in 2030?," *Slate*, October 20, 2011. www.slate.com.

Torie Bosch is the editor of *Future Tense*, an online publication sponsored by Arizona State University that reports on emerging technologies.

The California-based Post Carbon Institute commissioned a study to determine whether the earth's energy resources could sustain the industrial growth expected to occur over the remainder of the century. Among the findings of the 2009 report was this conclusion: Given global energy needs, renewable energy resources are insufficient to replace fossil fuels.

The report examined eighteen sources of energy—half of which are renewable—and concluded that renewable energy is not the answer to the earth's energy needs. The report found each source of renewable energy limited in its ability to provide reliable and sufficient power to meet the needs of cities, factories, and modes of transportation through the end of the twenty-first century.

Wind power, for example, is too unreliable. When breezes die down and wind turbines stop spinning, energy stops flowing. The Post Carbon Institute points out that Great Britain has invested in offshore wind turbines that produce energy from ocean breezes that blow toward the British Isles. But in 2009 a giant high-pressure system stalled over the British Isles. During a high-pressure system air is compressed in a small region, resulting in little wind. With little wind to power them, the wind turbines essentially ceased producing power.

Moreover, the high-pressure system occurred during a cold snap. This meant people turned up the heat in their homes at a time when the wind turbines were unable to provide much energy—meaning that coal-fired and nuclear power plants provided most of the energy during the cold snap. "Adding new wind generating capacity often does not substantially decrease the need for coal, gas or nuclear power plants; it merely enables those conventional power plants to be used less while the wind is blowing,"[22] the report said.

Limited Potential of Solar

The Post Carbon Institute also cited the limited potential of solar power to provide energy to an industrialized society. The report pointed out the enormous amount of energy available from the sun, finding that if just 1/40 of 1 percent of the sunlight that strikes the earth can be captured and converted to energy, it would be sufficient to provide all of the planet's electricity.

Making that happen would take an unprecedented dedication by world civilization to solar energy. Thousands of square miles would have to be made available for solar panel installations. And the Post Carbon Institute questioned whether such a commitment would be environmentally friendly, noting that great regions of forests and grasslands would have to be cleared to make way for the solar arrays, denying those habitats to native species. Said the Post Carbon Institute report, "It is not possible to point to more than a very few examples of an entire modern industrial nation obtaining the bulk of its energy from sources other than oil, coal, and natural gas."[23]

High Sticker Prices

One of the major roadblocks to weaning human society off fossil fuels is the huge task of replacing the existing energy infrastructure with renewable energy–production facilities. While it may take the dedication of thousands of square miles of open land to establish solar installations, as well as the erection of thousands of offshore wind turbines facing coastlines or standing upon prairie lands, those stationary facilities would not address the transportation needs of modern industrialized nations.

There are, in fact, more than 1 billion automobiles in the world, and the vast majority of them are propelled by gasoline. According to the North Carolina–based environmental group Think Global Green, just 2 percent are powered by electric motors.

Replacing some 998 million gas-powered cars with electric vehicles is a task few experts think can be accomplished. By 2013, although electric cars had gone into mass production, their sales were still widely outpaced by gasoline-powered vehicles. In America the two most popular electric cars—the Chevrolet Volt and Nissan Leaf—racked up sales of just some 33,000 vehicles in 2012. In that same year, Americans bought 13 million new gasoline-powered cars.

A problem is the price. The Volt and Leaf are both compact cars, yet the high cost of production—mostly in the expense of producing long-lasting batteries—has forced automakers to charge high prices for them. The sticker price for the Leaf in 2013 was nearly $30,000; the Volt's sticker price was nearly $40,000. Buyers can obtain similarly sized gas-powered cars for $10,000 to $20,000 less. Says Alan Boyle, science editor for NBC News, "As long as the cost of onboard electric power is high, compared to the cost of gasoline, buying an [electric vehicle] will never make sense based on fuel savings alone."[24]

Some experts believe it will take well into the 2020s before electric cars are affordable to most Americans. "I think we will see better batteries," says Jack Nerad, market analyst for Kelley Blue Book, which advises consumers on the values of cars. "Then electric cars become very, very viable. [For now] nothing is imminent."[25]

> "It is not possible to point to more than a very few examples of an entire modern industrial nation obtaining the bulk of its energy from sources other than oil, coal, and natural gas."[23]
>
> —Post Carbon Institute, a think tank that focuses on issues concerning energy, economics, and climate change.

Electric Car Batteries Are No Match for Gasoline-Fueled Engines

Electric car batteries have low "energy density," meaning they do not provide a lot of power and have to be recharged often, usually with electricity manufactured in coal-fired plants. The chart shows that the lithium-ion battery, the standard battery used in electric cars, produces a fraction of the energy density of gasoline, as well as other fuels. The chart measures the ability of a specific fuel to deliver a watt of power per kilogram of weight over the course of an hour. It takes 60 watts of power to illuminate an ordinary household lightbulb.

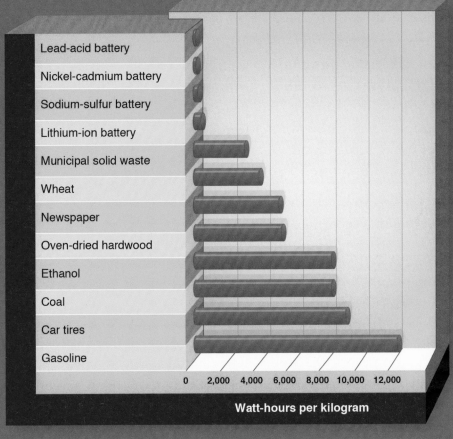

Note: The chart includes items such as newspapers and car tires—not normally considered to be fuel—to illustrate that even these items, when burned, have more energy density than the batteries used in electric cars.

Source: Robert Bryce, *Power Hungry: The Myths of Green Energy and the Real Fuels of the Future*. New York: PublicAffairs, 2010, p. 191.

Low Energy Density

Electric cars also have what is known as low energy density: An electric battery provides a fraction of the energy a gallon of gasoline can provide. In fact, all fossil fuels have high energy densities, particularly when compared with renewable energy sources. It takes a small amount of natural gas to make the same amount of energy produced by a solar array that covers thousands of square feet of land or rooftops. According to *Energy Tribune* editor Robert Bryce, natural gas has an energy density nearly five times the energy density of solar power and twenty-three times the energy density of wind.

> "As long as the cost of onboard electric power is high, compared to the cost of gasoline, buying an [electric vehicle] will never make sense based on fuel savings alone."[24]
>
> —Alan Boyle, science editor for NBC News.

As for electric cars, Bryce points out that to recharge an electric car, the owner must plug it into the house current for long periods—usually overnight. It is likely the electricity for the owner's house is provided by a coal-fired electric plant—energy that would not be needed if the car burned fossil fuels.

Moreover, skeptics argue that electric cars are indirectly responsible for releasing carbon emissions into the atmosphere. The batteries for these cars rely on the metal lithium as the conductive element that transfers the electricity from the battery to the engine. Mining lithium takes a lot of heavy equipment—backhoes and similar huge machines that are operated with diesel fuel. A study by the *Journal of Industrial Ecology* concluded that by the time an electric car powered by a lithium battery rolls out of the factory, the manufacturing process for that car has already produced 30,000 pounds (13,608 kg) of carbon emissions. In contrast, the manufacturing process for a conventional car produces 14,000 pounds (6,350 kg) in carbon emissions. Says Bjørn Lomborg, director of the Copenhagen Consensus Center in Washington, DC, and a well-known skeptic of the theory that fossil fuels are pollutants: "If a typical electric car is driven 50,000 miles over its lifetime, the huge initial emissions from its manufacture means the car will actually have put more carbon dioxide in the atmosphere than a similar-size gasoline-powered car driven the same number of miles."[26]

Irreplaceable Fuels

While automakers may be working toward production of affordable electric cars, the possibility of propelling other modes of transportation with renewable resources is still decades away—if, indeed, it is possible at all. Tractor-trailers—vital to the economies of industrialized nations—are powered by diesel fuel, which is oil based. According to the magazine *Popular Mechanics*, 68 percent of all goods shipped in America are transported by tractor-trailers, also known as semis. Currently, the technology to use electric power to fuel a huge tractor-trailer—which could weigh 20 tons (18 metric tons) or more—is in its infancy. A 2011 study by the Canadian environmental research organization FPInnovations found that it may be feasible to operate a hybrid tractor-trailer—meaning the semi would be powered by both an electric motor and a diesel-burning internal combustion engine. By 2013 it was not believed that any truck manufacturer was actively pursuing development of an all-electric semi capable of hauling a heavy, cargo-laden trailer.

Solar energy, wind power, and other forms of renewable energy are widely available, but each source of renewable energy is limited in its ability to meet the needs of industrialized societies. Coal, oil, and natural gas remain essential and largely irreplaceable in their abilities to fuel the engines that drive society in the twenty-first century and beyond.

Alternative Energy Can Take the Place of Fossil Fuels

"The 'fuel' for solar and wind is effectively limitless. For example, more potentially usable energy is received by the Earth from sunlight each and every hour than would be needed for all of the world's energy consumption in a full year. The potential for wind energy also exceeds the world's total energy demand several times over."

—Al Gore, *The Future: Six Drivers of Global Change*. New York: Random House, 2013, p. 282.

Al Gore is a former US vice president and a green energy advocate.

In 2013 the Solar Impulse airplane completed a flight from San Francisco, California, to New York City. It took two months for the propeller-driven plane to fly cross-country. During its coast-to-coast trip, the plane covered more than 2,500 miles (4,023 km) in five legs, making stops in Phoenix, Dallas–Fort Worth, Cincinnati, and Washington, DC, until finally arriving in New York on July 6. Average speed of the plane: 43 miles per hour (69 kph). Although it sounds as though the flight of the Solar Impulse is nothing special (a transcontinental flight aboard a jetliner takes about five hours), the plane covered the distance powered entirely by solar energy. The plane burned no aviation fuel, which is refined from oil.

The Solar Impulse received an enthusiastic welcome in New York as hundreds of people attended the landing, cheering as the plane came to a rest on the tarmac of John F. Kennedy International Airport. Among the luminaries who attended the landing were Erik Lindbergh, whose grandfather Charles made history in 1927 when he flew nonstop across the Atlantic Ocean; and James Cameron, director of the movie *Avatar* and other blockbuster films. "What Solar Impulse stands for is renewable energy—not just electric aircraft, but use of solar power in general, and this is something that's going to be fundamental and critical to the survival of the human race," Cameron said. "You've got people that are standing for something, committing themselves . . . to make a point for the betterment of human civilization, and I greatly applaud that."[27]

Rechargeable Batteries

According to Bertrand Piccard, also a Solar Impulse pilot, the main achievement of the aircraft was not its successful coast-to-coast flight but, rather, its employment of technology developed to make fuller use of solar energy than had been accomplished in the past. For example, even though the plane was powered by solar energy, it was still able to fly at night because batteries on board stored electricity collected through solar panels mounted atop the plane's wings. In most current home and industrial uses, solar energy is generally unavailable at night because there are few effective and low-cost ways to capture and store the power made during the day. But Solar Impulse engineers solved that problem by creating relatively lightweight rechargeable batteries—each of the plane's four batteries weighs 880 pounds (399 kg). Therefore, the batteries were light enough to enable the plane to maintain altitude.

Piccard suggests the technology used to power the plane could be applied to other modes of transportation, as well as residential and industrial uses—lighter batteries could be used in cars, for example. Also, homes with rooftop solar arrays could store their energy by using the battery technology tested by the Solar Impulse. "We are demonstrating new solutions to find renewable sources," says Piccard. "The really important technology on the airplane could already divide by two the energy consumption of the world and

"What Solar Impulse stands for is renewable energy—not just electric aircraft, but use of solar power in general, and this is something that's going to be fundamental and critical to the survival of the human race."[27]

—Film director James Cameron.

Conservation and Renewable Resources Could Eliminate the Need for Fossil Fuels

Even though demand for energy in New York is expected to rise through 2050, environmentalists believe conservation, as well as greater use of renewable resources, can over time eliminate the need for fossil fuels. This projection suggests demand can be reduced by increased use of "WWS" sources, which are identified as wind, water, and sun—meaning wind turbines, hydroelectric power plants, and solar energy. In addition, by conserving energy the need for fossil fuels can be eliminated by 2050.

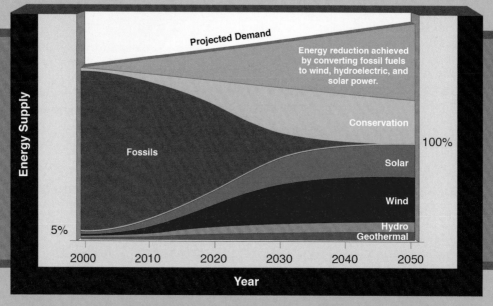

Source: Mark Fischetti, "How to Power the World Without Fossil Fuels," *Scientific American*, April 15, 2013. www.scientificamerican.com.

produce half of the needs of the renewable sources with the systems used on the airplane."[28]

Making Renewable Energy Viable

Some very influential people share Piccard's outlook. Since serving two terms as vice president, Al Gore has become a vocal advocate for replacing fossil fuels with renewable energy. Gore, who won a Nobel Peace Prize for his environmental activism, says, "We will have the rare privilege of meeting and overcoming the challenge that is worthy of the best in us. We have the tools we need. . . . What we most need is a shift in our thinking and a rejection of the toxic illusions that have been so assiduously promoted and continually reinforced by opponents of actions, principally large carbon polluters and their allies."[29]

Experts such as Gore acknowledge that the infrastructure to power cities, industries, and transportation through renewable resources is not yet in place, but they also point to innovations that could make alternative energy a much more viable source of power in the future. The flight of the Solar Impulse is one example. Indeed, NASA has developed its own solar-powered airplane, the *Pathfinder*, which the agency predicts could be a prototype of a new generation of aircraft that would not be propelled by petroleum-based jet fuel, but by the energy of the sun. Says a NASA statement, "The Pathfinder is a lightweight, solar-powered . . . flying wing aircraft that is demonstrating the technology of applying solar power for long-duration, high-altitude flight. It is literally the pathfinder for a future fleet of solar-powered aircraft."[30]

Free of Fossil Fuels

Some countries have already set goals of becoming 100 percent free of fossil fuels. Scotland has set a national goal of being 100 percent fueled by renewable energy by 2020. Two small Pacific Ocean nations, Tonga and the Cook Islands, have invested heavily in solar energy, hoping soon to be free of fossil fuels as well.

And countries in the Middle East, among them Saudi Arabia and the United Arab Emirates, have set similar goals. If these countries are able to cease powering their homes and businesses with fossil fuels it would be something of an ironic twist for a region that supplies a significant portion of the world's oil. Nevertheless, the people of these desert nations live under almost constant sunlight; therefore, their leaders see great promise in converting to solar energy. By 2013, Saudi Arabia and the Emirates each announced projects to erect solar arrays capable of providing energy to about 200,000 homes and businesses with plans to continue adding solar capacity through the end of the decade.

Tokelau, a territory of New Zealand composed of three small islands, has already achieved its goal. Since 2012 the territory has been able to power its homes and businesses entirely with renewable energy. Today solar panels dot the landscape of the islands. Prior to their installation, Tokelau imported and burned about two thousand barrels of diesel fuel a year to run electric generators. Says Mike Bassett-Smith, managing director of PowerSmart, the New Zealand solar energy company that installed the Tokelau panels, "All across the Pacific there are clear issues with the current and expected future costs of electricity generated using diesel, not to mention the environmental costs and risks of unloading diesel drums on tropical atolls. . . . Energy costs underpin the economic and social development of these nations."[31]

> "What we most need is a shift in our thinking and a rejection of the toxic illusions that have been so assiduously promoted and continually reinforced by opponents of actions, principally large carbon polluters and their allies."[29]
>
> —Al Gore, former vice president and a green energy advocate.

Commitment to Replacing Fossil Fuels

Tokelau is hardly an industrialized nation—fewer than fifteen hundred people live there, and exportation of handcrafts is the territory's major form of commerce. Nevertheless, renewable energy experts point to Tokelau as an example of how a community can commit itself to ridding its culture of fossil fuels.

Renewable energy advocates believe other communities can conserve, gradually reducing their reliance on fossil fuels with the ultimate goal of replacing them entirely with renewable resources. In 2013 Mark Jacobson, a professor of civil and environmental engineering from Stanford University in California, and Mark Delucchi, a research scientist at the Institute of Transportation Studies at the University of California–Davis, authored an article for *Scientific American* magazine in which they concluded that New York State could reduce its reliance on fossil fuels by 37 percent. The state can achieve that goal by erecting land-based and offshore wind turbines, establishing some eight hundred solar panel installations on open fields, and making use of 5 million residential rooftop solar installations. Says Jacobson, "My career has always been based on trying to understand large-scale pollution and climate problems—with the goal of trying to solve them. This is the 'trying to solve them' part. If society is going to do it, at least we now know that it's technically and economically feasible. Whether it actually happens depends on political will."[32]

Around the globe there are many examples of how alternative energy has replaced fossil fuels. The flight of the Solar Impulse and development of the *Pathfinder* illustrate how solar energy may one day revolutionize aviation. Meanwhile, many political leaders have come to the realization that life without fossil fuels is possible. In places like Scotland and Saudi Arabia, national leaders are planning for the arrival of that day, while on Tokelau that day has already arrived.

Chapter Four

Should the Government Continue to Support Fossil Fuels as an Energy Source?

The Government Should Continue to Support Fossil Fuels as an Energy Source

Hydraulic fracturing provides energy for American consumers, but the technology could not have been developed without aid from the federal government. That type of support is vital for Americans to enjoy energy security—the guarantee that energy will always be available to fuel their cars, heat their homes, and power their businesses. Government support can also go a long way toward ensuring that fossil fuels do not contribute to global warming; air capture technology could scrub the atmosphere of carbon, but the technology is in its infancy and needs government aid to become viable.

The Debate

The Government Should Not Continue to Support Fossil Fuels as an Energy Source

Government aid for fossil fuel companies dates back decades to when federal and state governments encouraged the growth of the energy industry. Now that fossil fuel producers are among the biggest corporations in America, many political leaders and others are calling for an end to taxpayer-financed subsidies. They believe fossil fuel producers should pay their fair share of taxes, in part to repair the environmental damage caused by oil, gas, and coal. Meanwhile, many city governments believe they can support clean energy policies by divesting their municipal treasuries of investments in fossil fuel corporations.

The Government Should Continue to Support Fossil Fuels as an Energy Source

"The lesson from the shale gas history is that government investment in innovation can, over time, commercialize and deploy technologies that make yesterday's less-efficient, dirtier, and more expensive technologies obsolete."

—Alex Trembath, "US Government Role in Shale Gas Fracking History: An Overview and Response to Our Critics," *Breakthrough*, March 2, 2012. http://thebreakthrough.org.

Alex Trembath is an energy policy analyst at the Breakthrough Institute.

The notion that natural gas could be forced to the surface through hydraulic fracturing was first proposed in 1975. In those early years engineers struggled to find ways to withdraw the gas from shale rock. One early test conducted in Morgantown, West Virginia, used explosives in an attempt to free the gas from pockets in the shale. The test proved disastrous. The explosives blew the pipeline out of the well, rocketing it into the sky more than 600 feet (183 m).

Other early attempts at hydraulic fracturing produced similarly bad results. Although those early tests failed to find effective ways to free natural gas from shale rock, they were nevertheless financed with grants from the Energy Department. Indeed, if it had not been for government assistance, it is likely the technology that makes fracking possible would never have been developed. Geologist Dan Steward started helping fracking pioneer Mitchell Energy find shale deposits in the early 1980s—an effort that included many failed attempts. "There's not a lot of companies that would stay with something this long," says Steward. "Most companies would have given up."[33]

Since those early days hydraulic fracturing has developed into a reliable source of energy—thanks in no small part to help from the Energy Department, which provided some $100 million in research grants, and to the federal and state governments, which since 1980 have provided hydraulic fracturing companies with billions of dollars in tax breaks. That type of government support has revolutionized the fossil fuel industry and will help make America energy independent. Says Steward, "The government has to be involved, to some degree, with new technologies."[34]

Energy Security

Government aid in the development of hydraulic fracturing has meant that more energy is produced within American borders, which has helped cut the need to import fossil fuels from other countries. Therefore, government aid for fracking has helped enhance America's energy security.

Energy security has long been a major concern for American consumers. Energy security is the guarantee that there will always be natural gas and coal to heat homes and power industries, as well as gasoline to fuel cars. Proponents of fossil fuels believe that by backing oil-, gas-, and coal-development projects in America and in its allies, the federal government can help ensure the country will always enjoy energy security.

Energy security is a concern because America imports 45 percent of its fossil fuels—and a significant portion of that fuel comes from the Organization of the Petroleum Exporting Countries, or OPEC. In 2013 America imported 33.5 percent of its oil from OPEC, whose members include Algeria, Angola, Ecuador, Iran, Iraq, Kuwait, Libya, Nigeria, Qatar, Saudi Arabia, United Arab Emirates, and Venezuela.

Generally speaking, OPEC's members act with a united voice, setting a single price for the oil they export. Some OPEC members are openly hostile to America, among them Venezuela and Iran. This means that when the OPEC oil ministers gather to decide what to charge for the crude they export to America, political considerations can trump economic factors. "None of [the OPEC] nations share the American vision of a free society based on the ideals of constitutionally limited government; none of them are particularly friendly to the United States," says Steven Yates, a professor of philosophy at the University of South Carolina Upstate and Greenville Technical College, who often writes on energy-related issues. "Political reasons could cause any of them to restrict or curtail the oil they now sell us."[35]

The Keystone XL Pipeline

To enhance energy security, many experts believe the US government should support efforts to obtain fossil fuels from friendly nations. Among the nations with the most potential to provide fossil fuels to America is Canada, which has enormous reserves of what is known as oil sands or tar sands. Located mostly in Alberta, the oil is found in sandy soil. According to the Alberta provincial government, Alberta's deposits make it one of the richest oil suppliers in the world, with 168 billion barrels of reserves—third only to Saudi Arabia's 264 billion barrels and Venezuela's 211 billion barrels.

Key to delivering Canadian oil to the American marketplace is the expansion of the Keystone Pipeline. The pipeline currently connects the Alberta oil sand fields with the city of Cushing, Oklahoma, a major refining and transportation hub in the American oil industry. Since the refineries and transit facilities in Cushing cannot keep up with the supply, the Canadian oil company TransCanada has proposed a $7 billion expansion that would add 1,700 miles (2,736 km) to the pipeline. This expansion would extend the line to refineries in Texas as well as provide a new leg connecting the Alberta oil sands with refineries in Kansas.

But the Canadians cannot build the expansion, known as Keystone XL, without the approval of the US government. By mid-2013 President

"None of [the OPEC] nations share the American vision of a free society based on the ideals of constitutionally limited government; none of them are particularly friendly to the United States. Political reasons could cause any of them to restrict or curtail the oil they now sell us."[35]

—Steven Yates, a professor of philosophy at the University of South Carolina Upstate and Greenville Technical College.

Government Support for Fossil Fuels Enhances Energy Security

To enhance energy security, the US government must support projects such as the Keystone XL pipeline, which would enable Canadian suppliers to deliver oil to refineries in Texas and Kansas. Canada is one of five main oil suppliers to the United States—and the two countries enjoy friendly relations. This is not the case with some of the other major suppliers, including Venezuela (which has hostile relations with the United States) and Saudi Arabia (which is located in the volatile Persian Gulf region). Government support for North American oil would go a long way toward enhancing energy security.

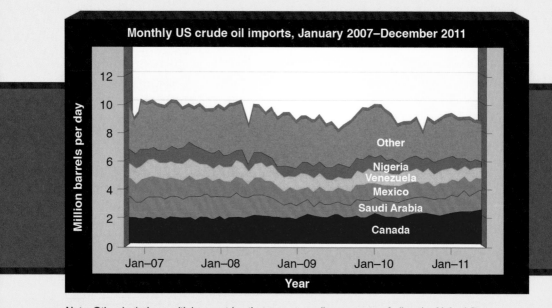

Monthly US crude oil imports, January 2007–December 2011

Note: Other includes multiple countries that export smaller amounts of oil to the United States.

Source: US Energy Information Administration, "Nearly 69 Percent of US Crude Oil Imports Originated From Five Countries in 2011," March 12, 2012. www.eia.gov.

Barack Obama had not yet granted that approval, due in large part to pressure from environmentalists who argue that the federal government should address climate change rather than find new ways to deliver fossil fuels to consumers. Russ Girling, the chief executive officer of TransCanada, argues that the Keystone Pipeline represents an important element in US energy security, jobs, and economic development. "If you deny the pipeline it's a lose-lose,"[36] says Girling.

Air Capture Technology

In 2013, as Obama continued to delay his decision on the Keystone Pipeline, he issued this challenge to the oil industry: "Allowing the Keystone Pipeline to be built requires a finding that doing so would be in our nation's interest. And our national interest will be served only if this project does not significantly exacerbate the problem of carbon pollution. . . . The net effects of climate impact will be absolutely critical to determining whether this project will go forward. It is relevant."[37]

Advocates of fossil fuels insist that there are ways in which oil from the Keystone Pipeline and other sources can be burned without adding carbon to the atmosphere—but it will take help from the government to make that happen. They have called on the federal government to provide research grants for technology that could scrub the air of carbons released by coal, oil, and natural gas. This process is known as air capture. In simple terms, air capture involves a widespread treatment of the atmosphere with chemicals that neutralize the carbon dioxide. In 2010 the Energy Department made $2.3 billion in grants available for researchers pursuing air capture technologies.

In an article written for the magazine *Atlantic*, Daniel Sarewitz, a professor of science and society at Arizona State University, and Roger Pielke Jr., a professor of environmental studies at the University of Colorado, point out that one of the major challenges of the research will be to develop methods for air capture that are inexpensive. Currently, it is estimated that it could cost $2,000 to remove a single ton of carbon from the atmosphere.

"Driving down the cost of carbon capture in the coming decades will require large government investments, not only for more research, but also to procure and deploy the technologies at scale. The climate is, of course, a public good; governments (and taxpayers) should fund a large part of any carbon-capture effort, just as they have funded other important public works."[38]

—Daniel Sarewitz, a professor of science and society at Arizona State University, and Roger Pielke Jr., a professor of environmental studies at the University of Colorado.

According to the Center for International Climate and Environmental Research in Oslo, Norway, each year the nations of the world pump more than 38 billion tons (34.5 billion metric tons) of carbon into the atmosphere. Currently, Sarewitz and Pielke acknowledge, air capture is not economically feasible; they call on the governments of the world to continue funding research. "Driving down the cost of carbon capture in the coming decades will require large government investments, not only for more research, but also to procure and deploy the technologies at scale," write Sarewitz and Pielke. "The climate is, of course, a public good; governments (and taxpayers) should fund a large part of any carbon-capture effort, just as they have funded other important public works."[38]

Backbone of the Economy

Fossil fuels are the backbone of the world economy—providing energy for consumers as well as industrial users. The government has played a role in their development, providing research grants and tax breaks to oil, coal, and gas companies to help them find new ways to deliver the energy. If fossil fuels are to remain an important component of the world economy, they will need government support to help energy companies continue finding new sources of coal, oil, and gas and making those fuels environmentally friendly.

The Government Should Not Continue to Support Fossil Fuels as an Energy Source

"Divesting from fossil fuels isn't just environmentally friendly, it's fiscally responsible. I think we want to make sure we're aligning our values and our concerns around climate change with our investment policies."

—Quoted in Neal J. Riley, "San Francisco May Divest Pension Funds in Oil Firms," *San Francisco Chronicle*, April 25, 2013. www.sfchronicle.com.

John Avalos is a San Francisco city supervisor.

The five largest oil companies doing business in America—BP, Chevron, Royal Dutch Shell, ConocoPhillips, and ExxonMobil—earned some $1 trillion in profits between 2001 and 2011. These profits were enhanced, to a large degree, by American taxpayers. Under a century-old law, oil companies are provided with tax breaks to help them find and produce oil. Each year, that tax break costs American taxpayers $4 billion.

Subsidies for oil companies have been a part of the American tax code since 1913. At the time, the environmental and health impacts of oil production were not evident, and Congress, anxious to aid a new and growing industry, provided oil companies with tax breaks. In those days oil companies were often small operations, many headed by scrappy roughnecks willing to brave inhospitable conditions in the American prairies and deserts to erect derricks. Today behemoth oil giants such as ExxonMobil and Chevron continue to enjoy subsidies. Under the current tax code, oil companies are permitted to deduct 15 percent of their operating costs from the money they pay in taxes.

Over the years, many Americans have called on Congress to end tax breaks for oil companies, but Congress has refused. Influential members of Congress from states where oil companies maintain drilling operations and refineries—including Texas and Alaska—have blocked efforts to repeal the tax breaks for oil companies. Alaska senator Mark Begich maintains that tax breaks for the oil companies may subsidize behemoths like ExxonMobil and Chevron, but they also benefit families because they help keep oil prices low. "Let's stop the headline grabbing and get serious about our developing a plan to secure our energy future," Begich says. "Getting rid of these incentives for domestic production won't decrease prices at the pump for our families and small businesses. Instead, it will discourage companies—especially the independents—from domestic investment and job creation."[39]

Nevertheless, many American political leaders, including President Obama, find it ludicrous that the federal government should be supporting industries that are able to rack up enormous profits. In 2012 Obama called for a repeal of the federal subsidies for oil companies. He said, "Every time you fill up your gas tank, they're making money. Does anyone really think Congress should give them another $4 billion this year? Of course not. It's outrageous. It's inexcusable. . . . Let's put every single member of Congress on record: You can stand with oil companies, or you can stand up for the American people."[40]

> "Every time you fill up your gas tank, [oil companies are] making money. Does anyone really think Congress should give them another $4 billion this year? Of course not. It's outrageous."[40]
>
> —Barack Obama, the forty-fourth president of the United States.

Tax Breaks for Frackers

Oil is not the only fossil fuel that enjoys tax breaks. Many state governments instituted tax breaks for hydraulic fracturing companies to help jump-start the industry, but now that fracking is an established industry, it is time to eliminate the breaks. In Texas, for example, lawmakers granted fracking companies tax incentives in 1989. Under Texas law, conventional natural gas producers pay a 7.5 percent tax on the gas they

Oil Companies Make Billions but Still Enjoy Huge Tax Breaks

The chart shows the net income—the profits made after expenses—for three oil companies: Shell, Chevron, and ExxonMobil. These companies made annual profits of tens of billions of dollars. Meanwhile, because of a century-old law, oil companies receive tax breaks that cost taxpayers some $4 billion a year.

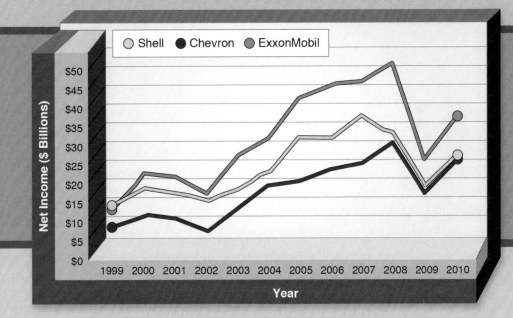

Source: Eli Rosenberg, "High Gas Prices Are Mostly Hurting You, Not Oil Companies," *The Atlantic*, May 13, 2011. www.theatlanticwire.com.

sell, but frackers can be taxed at rates as low as 2.5 percent. Now that the hydraulic fracturing industry makes enormous profits, some Texas lawmakers believe the frackers should pay the same tax as the conventional gas drillers. "[The tax break] definitely has to go," says Conor Kenny, an aide to Fort Worth legislator Lon Burnam. "There's no reason fracking needs incentives to survive."[41]

Moreover, in some states government officials say the economic benefits of the industry outweigh the risks to the environment and people's health. No place is this debate more heated than in Pennsylvania, where the state government has declined to tax hydraulic fracturing companies, believing that to do so would discourage them from establishing operations in the state. Governor Tom Corbett argues that fracking companies are a source of employment, but opponents insist fracking companies should be held accountable for repairing the damage to the environment and people's health by paying their fair share of taxes.

The governor eventually signed a bill enabling local communities to enact modest impact fees on hydraulic drillers—permitting those towns to recoup the cost of maintaining roads and other infrastructure damaged by the heavy vehicles and other equipment used in fracking. Critics suggested, though, that the fees do not come close to covering the environmental, health, and aesthetic costs of gas extraction. Under the terms of a hydraulic fracturing tax that was proposed by members of the Pennsylvania legislature in 2012, the state could earn nearly $50 billion through 2032 by taxing the gas producers. However, that bill died in the legislature when Corbett made it clear he would not sign it. "Wealthy drillers are making tons of money while dramatically changing the character of the countryside—and they ought to pay their fair share," insists Greg Vitali, a Pennsylvania state legislator who supports a tax on hydraulic fracturing companies. "It's absolute insanity. It's absolutely indefensible."[42]

Divesting Oil Stocks

Government officials concerned with greenhouse gas emissions and other ill effects of fossil fuels have found a way to withdraw their support from oil, gas, and coal corporations. They are dropping, or divesting, their communities' investments in energy companies. Over the years, many cities and towns have invested their assets—usually pension funds—in the publicly traded stocks of fossil fuel companies. Leaders of some cities believe that if enough American communities join together and divest their assets of fossil fuel stocks—and instead invest them in clean energy companies—fossil fuel companies could be economically crippled, leading to the growth of the clean energy field.

One city that divested its assets of fossil fuel stocks is Richmond, California, home to a Chevron refinery. Mayor Gayle McLaughlin says the city eliminated its fossil fuel investments to send a message to the industry that it is no longer welcome in Richmond. "Richmond is home to the second largest oil refinery and largest point source of greenhouse gas emissions in California," says McLaughlin. "I am proud to join with other cities in this divestment campaign, as we divest from an industry that is wreaking havoc on our community and planet, and reinvest in a clean energy economy with new jobs for our residents."[43]

By 2013 other cities that had taken steps to divest their municipal treasuries of fossil fuel stocks were Madison and Bayfield, Wisconsin; Ithaca, New York; Boulder, Colorado; Rochester, Minnesota; Eugene, Oregon; Berkeley and San Francisco, California; and Seattle, Washington. In Seattle, Mayor Mike McGinn empaneled a commission in 2012 to develop a plan aimed at weaning the city entirely off fossil fuels by 2050. McGinn set that goal because he feels city governments should be part of the worldwide effort to combat climate change.

> "I am proud to join with other cities in this divestment campaign, as we divest from an industry that is wreaking havoc on our community and planet, and reinvest in a clean energy economy with new jobs for our residents."[42]
>
> —Mayor Gayle McLaughlin of Richmond, California.

Part of the plan includes ensuring that the city does not invest its assets in fossil fuel companies. According to McGinn, the city's $2.6 billion pension fund has included investments in fossil fuel companies, and he called on the plan's administrators to sell those assets. Says McGinn, "Divestment is just one of the steps we can take to address the climate crisis. . . . Cities that [divest] will be leaders in creating a new model for quality of life, environmental sustainability and economic success. We've got a head start on that here in Seattle, but there's a lot more work to do."[44]

Facing Powerful Foes

As the actions taken by Richmond and Seattle illustrate, some government leaders do not believe government should use taxpayer money to support the fossil fuel industry. In Washington, DC, many political leaders have called for an end to the subsidies that aid such oil giants as Exxon-Mobil, Royal Dutch Shell and ConocoPhillips. On the state level, many lawmakers have demanded an end to the government aid that has helped establish and support the fracking industry. They argue that hydraulic fracturing is now a healthy industry that should be made to pay its fair share in taxes—largely to help repair the damage that fracking causes to the environment. Government support for the fossil fuel industry has a long history, though, and advocates for ending government aid to oil, gas, and coal interests realize they face powerful foes who are not ready to give up the benefits they have come to expect from taxpayers.

Source Notes

Overview: Visions of the Future: Fossil Fuels

1. Quoted in Karl Vick, "Tapping the Promised Land: Can Israel Be an Energy Giant?," *Time*, April 30, 2013. http://world.time.com.
2. Robert Bryce, *Power Hungry: The Myths of Green Energy and the Real Fuels of the Future*. New York: PublicAffairs, 2010, p. 14.
3. Sascha Müller-Kraenner, *Energy Security*. London: Earthscan, 2007, p. 3.

Chapter One: Are Fossil Fuels Affordable?

4. Nigel Lawson, "Thought We Were Running Out of Fossil Fuels? New Technology Means Britain and the US Could Tap Undreamed Reserves of Gas and Oil," *Daily Mail*, December 7, 2012. www.dailymail.co.uk.
5. Quoted in Steve Hargreaves, "Why Cheap Oil Is Here to Stay," CNN, December 3, 2009. http://money.cnn.com.
6. Bryce, *Power Hungry*, p. 242.
7. Thanassis Cambanis, "American Energy Independence: The Great Shake-Up," *Boston Globe*, May 26, 2013. www.bostonglobe.com.
8. Quoted in Amy Sinatra Ayres, "Isaac Drives Spike in U.S. Gas Prices Ahead of Labor Day Weekend," National Geographic News, August 30, 2012. http://news.nationalgeographic.com.
9. Quoted in Paul Brown, "Arctic Oil and Gas Will Exact High Price," Climate News Network, March 6, 2013. www.climatenewsnetwork.net.
10. Quoted in Robin Madel, "Frackonomics: Debunking the Financial Myths of Fracking," *Huffington Post*, June 12, 2012. www.huffingtonpost.com.
11. Quoted in Jennifer Peltz, "Levees, Floodwalls Urged for New York," *Philadelphia Inquirer*, June 12, 2013, p. A-4.

Chapter Two: Can Fossil Fuels Be Compatible with the Environment?

12. Quoted in Erin Ailworth, "Shell's CEO Says Gas Will Lead the Way," *Boston Globe*, March 22, 2013. www.bostonglobe.com.

13. Quoted in ScienceDaily, "Carbon Capture and Storage to Combat Global Warming Examined," June 12, 2007. www.sciencedaily.com.

14. Quoted in S. George Philander, ed., *Encyclopedia of Global Warming and Climate Change*. Thousand Oaks, CA: Sage, 2008, p. 216.

15. Quoted in Australian Broadcasting Corporation, "Ian Plimer Discusses His Book Denying Global Warming," April 27, 2009. www.abc.net.au.

16. Quoted in James Delingpole, "Meet the Man Who Has Exposed the Great Climate Change Con Trick," *Spectator*, July 8, 2009. www.spectator.co.uk.

17. Quoted in Bill Dedman, "Disputes over Environmental Impact of 'Fracking' Obscure Its Future," NBC News, April 7, 2013. http://openchannel.nbcnews.com.

18. Quoted in Dedman, "Disputes over Environmental Impact of 'Fracking' Obscure Its Future."

19. National Wildlife Federation, "How Does the BP Oil Spill Impact Wildlife and Habitat?," 2013. www.nwf.org.

20. Quoted in National Aeronautics and Space Administration, "NASA Finds 2012 Sustained Long-Term Climate Warming Trend," January 15, 2013. www.nasa.gov.

21. Quoted in National Aeronautics and Space Administration, "Global Climate Change: Vital Signs of the Planet," 2012. http://climate.nasa.gov.

Chapter Three: Can Alternative Energy Take the Place of Fossil Fuels?

22. Richard Heinberg, *Searching for a Miracle: 'Net Energy' Limits & the Fate of Industrial Society*. Santa Rosa, CA: Post Carbon Institute, 2009. www.postcarbon.org.

23. Heinberg, *Searching for a Miracle*.

24. Alan Boyle, "The Race to Replace Gasoline," NBC News, February 11, 2011. http://cosmiclog.nbcnews.com.

25. Quoted in Alex Davis, "The Electric Car Is Dead All Over Again," Business Insider, April 8, 2013. www.businessinsider.com.

26. Bjørn Lomborg, "Green Cars Have a Dirty Little Secret," *Wall Street Journal*, March 11, 2013. http://online.wsj.com.

27. Quoted in Alan Boyle, "Solar Impulse Plane Ends American Odyssey with Fears, Tears and Cheers," NBC News, July 6, 2013. www.nbcnews.com.

28. Quoted in Alan Hayley Ringle, "Solar Impulse Plane Makes Stop in Phoenix," *Phoenix Business Journal*, May 7, 2013. www.bizjournals.com.

29. Al Gore, *The Future: Six Drivers of Global Change*. New York: Random House, 2013, p. 358.

30. National Aeronautics and Space Administration Dryden Flight Research Center, "*Pathfinder* Solar-Powered Aircraft," November 2002. http://web.archive.org.

31. Quoted in Paul E. McGinniss, "Island of Tokelau Becomes World's First Solar-Powered Country," EcoWatch, November 15, 2012. http://ecowatch.com.

32. Quoted in Mark Fischetti, "How to Power the World Without Fossil Fuels," *Scientific American*, April 15, 2013. www.scientificamerican.com.

Chapter Four: Should the Government Continue to Support Fossil Fuels as an Energy Source?

33. Quoted in *Cleveland Plain Dealer*, "Tax Breaks, US Research Play Big Part in Success of Fracking," September 24, 2012. www.cleveland.com.

34. Quoted in *Cleveland Plain Dealer*, "Tax Breaks, US Research Play Big Part in Success of Fracking."

35. Steven Yates, "Why So High?," *New American*, July 7, 2008, p. 13.

36. Quoted in Zack Colman, "Keystone Builder 'Extremely Confident' Obama Will Approve It," *The Hill* (blog), May 31, 2013. http://thehill.com.

37. Quoted in Brad Plumer, "Obama May Have Left Himself Wiggle Room to Approve Keystone XL," *Wonkblog, Washington Post*, June 25, 2013. www.washingtonpost.com.

38. Daniel Sarewitz and Roger Pielke Jr., "Learning to Live with Fossil Fuels," *Atlantic*, May 2013, p. 59.

39. Quoted in Amanda Coyne, "Begich, Murkowski Oppose Linking Deficit, Oil 'Subsidies,'" Alaska Dispatch, May 11, 2011. www.alaskadispatch.com.

40. Quoted in David Nakamura, "Obama Calls on Congress to Repeal Federal Subsidies for Oil Industry," *Washington Post*, March 1, 2012. www.washingtonpost.com.

41. Quoted in Jesse Davis, "Texas Dems Seek to End State's Fracking Tax Breaks," Law 360, November 13, 2012. www.law360.com.

42. Quoted in Will Bunch, "How Corbett Fracked Pa's Middle Class," *Philadelphia Daily News*, March 8, 2011. www.philly.com.

43. Quoted in SustainableBusiness.com, "10 Cities Divest from Fossil Fuel Investments," April 26, 2013. www.sustainablebusiness.com.

44. Mike McGinn, "An Update on Fossil Fuel Divestment," Mayor Mike McGinn, December 21, 2012. http://mayormcginn.seattle.gov.

Fossil Fuel Facts

Production

- Frackers extract gas relatively close to the surface—most wells are no more than 5,000 feet (1,524 m) deep; however, hydraulic fracturing wells are also drilled horizontally, some extending for 1 mile (1.6 km) or more.
- The world's deepest oil well is owned by a subsidiary of ExxonMobil. Located in Chayvo, Russia, the well draws oil from 7.7 miles (12.4 km) beneath the earth's surface.
- The northernmost oil rig in the world is owned by the Norwegian oil company Statoil; the rig, known as *West Hercules*, is located in the Barents Sea about 1,200 miles (1,930 km) south of the North Pole.
- OPEC reported in 2013 that its twelve member nations produce 30 million barrels of crude oil a day.
- According to the Energy Information Administration, each year American mines produce about 1 million tons (907,185 metric tons) of coal. Coal is mined in twenty-five states, with Wyoming, West Virginia, Kentucky, Pennsylvania, and Texas producing the most.
- The Energy Information Administration reported in 2013 that Texas produces the most natural gas among the states, with an annual production of 7.1 trillion cubic feet (201 billion cu. m). Other top producers among the states are Louisiana, 3 trillion cubic feet (85 billion cu. m); Wyoming, 2.2 trillion cubic feet (62 billion cu. m); Oklahoma, 1.9 trillion cubic feet (54 billion cu. m); and Colorado, 1.6 trillion cubic feet (45 billion cu. m).

Consumption

- According to the advocacy group American Energy Independence, each day American automobiles burn through 378 million gallons (1.4 billion L) of gasoline.

- The advocacy group Union of Concerned Scientists reports that the typical coal-fired electricity-generating plant in America burns 1.4 million tons (1.3 million metric tons) of coal each year. As of 2012 there were 572 such plants operating in America.
- The Energy Information Administration reports consumption of jet fuel has risen steadily since the 1980s; in 1984 the worldwide consumption of jet fuel stood at 1.8 million barrels; in 2010 jet aircraft burned through 5.2 million barrels.
- Oil and gas are routinely imported and exported, but coal is usually used in the country where it is mined; the Center for Climate and Energy Solutions reported in 2013 that only 16 percent of coal mined worldwide is exported to other countries.

Cost

- In the United States in 2013, gasoline cost an average of $3.60 a gallon; only about 67 percent of that cost was devoted to paying for the crude oil refined into gasoline. Other costs include 14 percent for refining, 12 percent for highway taxes, and the remainder for distribution and marketing.
- According to 2010 statistics compiled by the Air Transport Association, nearly 20 percent of the cost of an airline ticket is devoted to the cost of fuel. A passenger holding a $506 ticket for a cross-country flight would have paid about $98 in fuel costs.
- The World Coal Association reported in 2011 that it costs as little as $56 to produce 1 megawatt of electricity by using coal, whereas it can cost as much as $90 to produce 1 megawatt by using nuclear fuel. One megawatt can produce enough energy to meet the needs of about six hundred homes for a year.
- According to the Energy Department, a gallon of gasoline cost 25 cents in 1919. When adjusted for inflation, that same gallon of gas would have cost $3.75 in 2011. In reality, though, American consumers paid $3.48 per gallon in 2011.
- Despite the widespread availability of natural gas through hydraulic fracturing, consumer prices have risen over the past three decades. The Energy Information Administration reports the price of gas jumped

from $3.94 per thousand cubic feet of gas in 1981 to $9.24 per thousand cubic feet in 2013.

Fossil Fuels and the Environment

- A 2013 report by the Environmental Protection Agency found that the generation of electricity produces 33 percent of greenhouse gases, and the use of cars, trucks, and airplanes accounts for 28 percent of greenhouse gas emissions. Other producers of greenhouse gases are industrial, 20 percent; commercial and residential, 11 percent; and agricultural production, 8 percent.
- According to statistics released in 2011 by the Energy Information Administration, China is the planet's top emitter of greenhouse gases, producing about 6.8 billion tons (6.2 billion metric tons) per year. America is the next top producer, accounting for some 5.8 billion tons (5.3 billion metric tons).
- The Energy Department reported that in 2013 the most fuel-efficient gasoline-powered cars were the Scion iQ and Honda CR-Z, both of which achieved a fuel economy of 37 miles per gallon (15.7 km/L); the least fuel-efficient car was the Bugatti Veyron, a sports car that achieved a fuel economy of 10 miles per gallon (4.3 km/L).
- According to the Union of Concerned Scientists, the typical coal-fired electricity plant in America each year emits 10,000 tons (9,072 metric tons) of sulfur dioxide, a cause of acid rain; 10,200 tons (9,253 metric tons) of nitrogen oxide, a cause of smog and acid rain; and 3.7 million tons (3.4 million metric tons) of carbon dioxide, a greenhouse gas.
- Most states require fracking companies to disclose the chemicals they use in their extraction processes, but in 2012 an investigation by the publication *EnergyWire* found that in Pennsylvania, 65 percent of those disclosures omit information about the chemicals used by frackers.

Fossil Fuels and the Role of Government

- The advocacy group Center for American Progress reported in 2012 that for every dollar that oil companies spend on lobbying in Washington, DC, they earn thirty dollars in tax breaks.

- The Energy Department has placed a monetary figure on the damage caused by the emission of a single ton of carbon: In 2010 the department set that figure at twenty-two dollars; in 2013 the agency adjusted the figure upward to thirty-six dollars.
- The American Petroleum Institute reported in 2011 that oil companies pay $85 million a day in federal taxes.
- The Russian government announced in 2013 that it would petition the United Nations to recognize new borders for the country that would include 745,000 square miles (1.9 million sq. km) of Arctic territory; Russia wants recognition of the new borders in order to lay claim to new sources of oil in the Arctic.

Related Organizations and Websites

American Petroleum Institute (API)
1220 L St. NW
Washington, DC 20005-4070
phone: (202) 682-8000
website: www.api.org

The national trade association representing the American oil and natural gas industries, the API produces publications ranging from technical manuals to guides for consumers on how to get the most value for the money they spend on energy. Visitors to the institute's website can find the API's position on climate change and other environmental issues involving fossil fuels.

Citizens Climate Lobby
1330 Orange Ave., No. 300
Coronado, CA 92118
phone: (619) 437-7142
e-mail: ccl@citizensclimatelobby.org
website: http://citizensclimatelobby.org

Citizens Climate Lobby organizes at the grassroots level with the goal of lobbying elected officials to adopt laws restricting the use of fossil fuels and promoting clean energy. Visitors to the organization's website can find information on a number of proposed laws supported by the group, many of which would place heavy taxes on commercial fossil fuel users.

National Mining Association (NMA)
101 Constitution Ave. NW, Suite 500E
Washington, DC 20001
phone: (202) 463-2600
website: www.nma.org

The NMA is the national trade group for the American coal mining industry. Visitors to the organization's website can find statistics on mining in America, including annual production and a region-by-region breakdown on electricity generated by burning coal. Copies of the fact sheets *Coal: America's Power* and *Economic Contributions of Mining* can be downloaded from the site.

National Wildlife Federation
11100 Wildlife Center Dr.
Reston, VA 20190
phone: (800) 822-9919
website: www.nwf.org

The environmental group National Wildlife Federation monitors the effects of pollution on wildlife. The organization also keeps close watch over how each of the fossil fuels contributes to global warming. By following the link on the organization's website for "Climate and Energy," students can find information on the environmental impacts of coal, oil, and natural gas.

Organization of the Petroleum Exporting Countries (OPEC)
Helferstorferstrasse 17
A-1010
Vienna, Austria
website: www.opec.og

OPEC is composed of twelve countries that export oil. Visitors to the OPEC website can find a history of the cartel and download copies of *World Oil Outlook*, which projects the amount of oil available for worldwide consumption. By following the link for "Data/Graphs," visitors can see the daily "basket price," which reflects the price for oil by the barrel.

Post Carbon Institute
613 Fourth St., Suite 208
Santa Rosa, CA 95404
phone: (707) 823-8700 • fax: (866) 797-5820
website: www.postcarbon.org

The Post Carbon Institute provides information to consumers and communities that desire to plan for a future free of fossil fuels. Visitors to the organization's website can find copies of the institute's report *Searching for a Miracle*, which examines the ability of renewable resources to meet worldwide energy needs.

TransCanada Corporation
450 1 St. SW
Calgary, AB, Canada T2P 5H1
phone: (800) 661-3805 • fax: (403) 920-2200
website: www.transcanada.com

TransCanada is the Canadian oil company that proposes to expand the Keystone Pipeline to deliver more crude from Alberta. The company maintains a separate website, http://keystone-xl.com, that explains the expansion. Students can find a link for "Energy Security" on the website that will explain how Americans could rely less on oil imported from unfriendly nations.

Union of Concerned Scientists
2 Brattle Sq.
Cambridge, MA 02138-3780
phone: (617) 547-5552 • fax: (617) 864-9405
website: www.ucsusa.org

Composed of more than four hundred thousand scientists and other members, the Union of Concerned Scientists provides analyses of issues the group believes endanger the planet's future, including global warming. Students can follow the link for "Extreme Heat and Climate Change" on the organization's website to find out more about this topic.

US Department of Energy
1000 Independence Ave. SW
Washington, DC 20585
phone: (202) 586-5000
website: www.energy.gov

The Energy Department is the federal government's chief regulatory agency over all energy used in the country. Visitors to the agency's website can find many resources about fossil fuel production and consumption. By following the link for "Energy Sources," students can find information on carbon capture technology and compressed natural gas vehicles.

US Environmental Protection Agency (EPA)
1200 Pennsylvania Ave. NW
Washington, DC 20460
phone: (202) 272-0167
website: www.epa.gov

The EPA was created by Congress to act as the federal government's chief watchdog over the environment, monitoring air and water quality. By following the link on the agency's website for "Science & Technology," students can find information on air quality, including how the EPA measures emissions from automobiles, coal-fired plants, and other consumers of fossil fuels.

For Further Research

Books

Robert Bryce, *Power Hungry: The Myths of Green Energy and the Real Fuels of the Future*. New York: PublicAffairs, 2010.

Leonardo Maugeri, *Beyond the Age of Oil: The Myths, Realities, and Future of Fossil Fuels and Their Alternatives*. Santa Barbara, CA: Praeger, 2010.

Bill McKibben, *The Global Warming Reader: A Century of Writing About Climate*. New York: Penguin, 2012.

Francisco Parra, *Oil Politics: A Modern History of Petroleum*. New York: I.B. Tauris, 2009.

Tom Wilber, *Under the Surface: Fracking, Fortunes, and the Fate of the Marcellus Shale*. Ithaca, NY: Cornell University Press, 2012.

Periodicals

James Fallows, "Dirty Coal, Clean Future," *Atlantic*, December 2010.

Jeff Goodell, "Obama's Climate Challenge," *Rolling Stone*, January 31, 2013.

Joshua Hammer, "The Land of Oil and Money," *Fast Company*, September 2011.

Charles C. Mann, "What If We Never Run Out of Oil?," *Atlantic*, May 2013.

Bryan Walsh, "The Future of Oil," *Time*, April 9, 2012.

Websites

Chemicals Used in Natural Gas Operations (www.endocrinedisrup
tion.com/chemicals.introduction.php). This website is maintained by
the Endocrine Disruption Exchange, a group that monitors chemicals
affecting the endocrine system, which consists of glands in humans and
other mammals. The site provides lists of chemicals and other resources
about the effects of fracking.

Hydraulic Fracturing Facts (www.hydraulicfracturing.com/Pages/infor
mation.aspx). Maintained by Chesapeake Energy, a natural gas producer,
this site explains the process of hydraulic fracturing. Students can watch
an online video showing how natural gas is forced to the surface through
hydraulic fracturing.

Israel Energy Initiatives (IEI) (www.iei-energy.com). The website of the
Jerusalem-based company explains how IEI intends to withdraw oil shale
from the Valley of Elah. Visitors to the website can find information on
the composition of oil shale and the technology involved in forcing the
fossil fuel to the surface.

Solar Impulse (www.solarimpulse.com). Visitors to the official website
of the solar-powered airplane can see photos and videos of the craft's
2013 cross-country flight, read an overview of the technology that has
gone into the plane, and access schematic drawings of the aircraft.

Tokelau Renewable Energy Project (http://powersmartsolar.co.nz
/blog/id/413). Maintained by PowerSmart, the New Zealand solar energy company, this website describes the company's project to make the
three-island territory free of fossil fuels. Visitors to the website can find
photos of solar panels on Tokelau as well as a description of the project.

Index

Note: Boldface page numbers indicate illustrations.